The CBC

The CBC

How Canada's Public Broadcaster Lost Its Voice (And How to Get It Back)

David Cayley

SUTHERLAND HOUSE

TORONTO, 2025

Sutherland House
416 Moore Ave., Suite 304
Toronto, ON M4G 1C9

First edition, September 2025

If you are interested in inviting one of our authors to a live event or media appearance, please contact sranasinghe@sutherlandhousebooks.com and visit our website at sutherlandhousebooks.com for more information.

We acknowledge the support of the Government of Canada.

Manufactured in China
Cover designed by Luisa Galstyan and Jordan Lunn

Library and Archives Canada Cataloguing in Publication
Title: The CBC : how Canada's public broadcaster lost its voice
(and how to get it back) / David Cayley.
Names: Cayley, David, author
Description: Includes bibliographical references and index.
Identifiers: Canadiana (print) 20250190079 | Canadiana (ebook) 20250190168 |
ISBN 9781998365548 (softcover) | ISBN 9781998365166 (EPUB)
Subjects: LCSH: Canadian Broadcasting Corporation—History. |
LCSH: Public broadcasting—Canada—History.
Classification: LCC HE8689.9.C3 C39 2025 | DDC 384.540971—dc23

ISBN 978-1-998365-54-8
eBook 978-1-998365-16-6

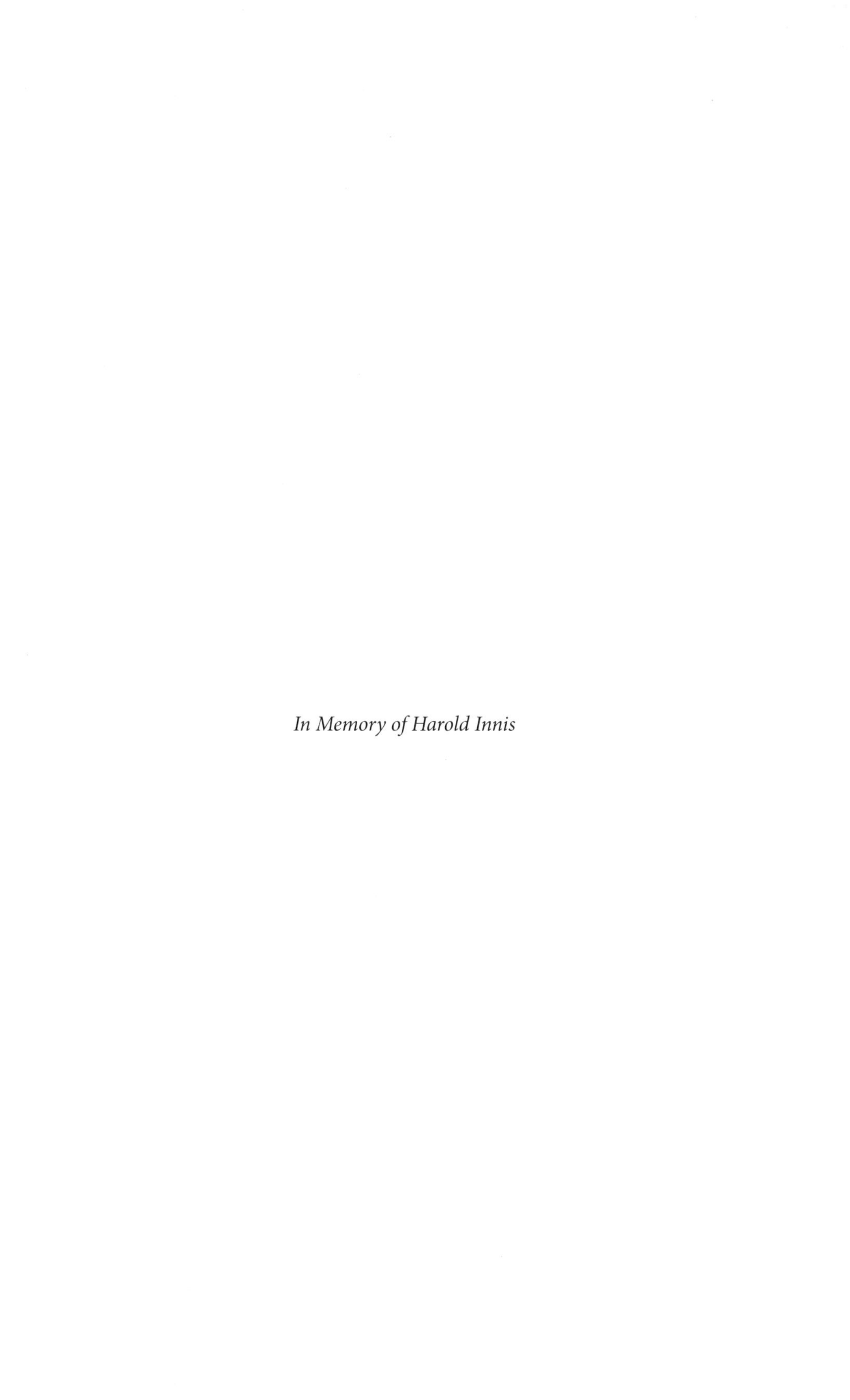

In Memory of Harold Innis

Contents

Preface

This book goes to press at a moment when the election campaign of 2025 is still in progress and its outcome still uncertain. The Conservative Party has pledged to "defund" the CBC, excepting *Radio Canada*, and its leader, Pierre Poilievre, has even threatened to turn the CBC's English language headquarters into a housing project. The Liberal Party has said it will increase funding to the CBC and "renew" its mandate. Both parties assume they know what the CBC is and what it's for. This book puts that assumption in question. It argues that political polarization, new media technologies, and various cultural confusions have rendered the very purpose of public broadcasting obscure, uncertain, and in need of redefinition. The CBC is at the end of an era, in its own life and in the larger life of its country, and will need to reorient itself for a new age. This fundamental task will remain, whether the CBC is being dismantled or reinforced by the time this book reaches its readers.

Introduction

In the fall of 1964, the producers of a national television newsmagazine called *This Hour Has Seven Days* started a revolution at the Canadian Broadcasting Corporation (CBC). According to the manifesto with which the program began its first broadcast, the new program would "probe hypocrisy," right "public wrongs," "grill ... prominent guests," and, in the process, create "journalism of ... such urgency that it will become mandatory viewing for a large segment of the nation."[1] The CBC's top executives in Ottawa were immediately alarmed.

Management, at the time, was mostly made up of radio veterans, who had come of age during CBC's long struggle to establish an independent, "arms-length" relationship with the government. This relationship, in their view, was premised on the CBC's political neutrality, and they feared that the independence the CBC had won on this basis would be compromised by the activism of the young TV generation that was beginning to flex its muscles at *Seven Days*. "The CBC was not brought into being to instigate or stimulate social change," said CBC President Alphonse Ouimet. It must, he said, "serve public opinion rather than moulding it," and allow "people to make up their own minds."[2]

Confrontation was inevitable, and the new show lasted only two stormy seasons. After three broadcasts, the vice president of programs, Eugene Hallman, set out head office's position in an unequivocal policy statement: "No program can be permitted to adopt an editorial point-of-view or take a position in matters of public controversy."[3] But *Seven Days* had the *zeitgeist*, an eager audience and the thrilling new possibilities of television on its side, and its staff were unwilling to back down. In the spring of 1966, the show was cancelled.

A public outcry followed: Alphonse Ouimet was hanged in effigy by an inflamed crowd outside the Vancouver Courthouse, and, in the end, Prime Minister Lester Pearson had to intervene to ensure the last episode of the show aired. Three million Canadians watched—an astonishing number for a public affairs program at a time when the Canadian population was barely twenty million.

Inside the CBC, this struggle for the corporation's soul was often called "the *Seven Days* war." It was a war in which *Seven Days* clearly won the ultimate victory, even if Ottawa was first able to exact its pound of flesh. In his history of these events, *Inside Seven Days*, former CBC producer and manager Eric Koch says that the affair brought an end to "the first era of [Canadian] broadcasting."[4] Koch was in a position to know. When the trouble started, he was a supervising producer in television current affairs, and he felt the struggle in his own soul. He was "greatly impressed," he writes in his book, "by the idealism, courage and vitality of the [*Seven Days*] producers," but he was also, in many ways, a man of what he called "the first era," having joined the CBC in 1943.[5] During this era, Koch says, "the CBC had many features of a secular church with the President as Pope and the Broadcasting Act as gospel." This church-like character was expressed in a strong *esprit de corps*, a sense of public service, and a certain formal reserve in the corporation's approach to its publics. Producer Bernard Trotter, a colleague of Koch's in the radio service's Department of Talks and Public Affairs, characterized this ethos as that of a public utility—a "free forum" on the model of "the Athenian *agora*," with the CBC as its active but neutral convener.[6] *Seven Days* was obviously different, and not just because its manners were brash and its journalism more aggressive than the CBC was used to. Its difference, above all, was in making the approval of its audience its *raison d'être* and its prime source of legitimacy. The CBC had always wanted to serve its audiences, but, in the past, it had done so according to standards that overrode popular acclaim. An example would be Alphonse Ouimet's belief that the corporation had no business trying to "instigate social change." *Seven Days* wanted to stir the people up. Its motto, in the words of its producer and co-host Patrick Watson, was: "We got to have them watching."[7]

Within the CBC, this struggle between philosophies was often said to pit populism against elitism. The terms are fairly crude idealizations—the CBC was sometimes "populist" before 1964, and sometimes "elitist" afterward—and the word "populism" has the additional disadvantage of being currently used for a movement on the political right, whereas, at that time, it evoked a perceived move to the left. But, with these caveats, I will retain the term. It was the word people used at the time, and I know of no better one-word description of the ethos that dominated the CBC in the years after the *Seven Days* revolution.

I think that Eric Koch was right in seeing the *Seven Days* war as the boundary between two distinct eras at the CBC. The program's claim that it could speak *for* its audience, as its delegate, substitute, and tribune, was a sign that a new era was beginning. The old radio hands who had torpedoed *Seven Days* soon disappeared from the management suites. The show that replaced *Seven Days* on Sunday evenings was, if anything, more uninhibited than its successor. Called simply *Sunday*, it offered what its director, Daryl Duke, called "psychedelic ... total television" with the atmosphere of "a medieval bearpit or a bullfight arena."[8] The same ethos soon spread to CBC Radio, where the so-called radio revolution of the late 1960s and 1970s reproduced all the most significant features of *Seven Days*' populism—its sense of mission, its identification with audiences, its pursuit of immediacy. As late as 2004, when Richard Stursberg became the head of the CBC's English radio and television networks, the revolution was still in progress. Stursberg had no sooner taken office than he set out to uproot the last traces of *elitism* at the CBC. His new colleagues, he claimed, were suffering "a profound schizophrenia," which made them believe that "if [a] show attracted large audiences, it must be vulgar and stupid."[9] "Whether Canadians watched," he said, was "the only relevant criterion" by which the success of a program could be judged because its audience "were the only judges who counted."[10]

It will be my claim in this book that the populist era inaugurated by *Seven Days* is now coming to an end. The public that populism wanted to serve has shattered. Patrick Watson, one of the creators of *Seven Days* and its co-host in its second season, says in his autobiography that he was taught by his mentor, CBC Television pioneer Ross McLean, to always ask of a program, "How will it serve the audience?"[11] *The audience* in this conception is singular and homogeneous, as is *the public* of which it forms a representative fraction. I don't mean to imply that McLean, or Watson, or anybody else who took this view, were simpletons, somehow unaware of the many, quite obvious fissures that then existed in Canadian society. I mean rather that they imagined their audience as an ideal community possessing a common interest that the CBC existed to identify and serve. I do not believe that this common character exists any longer, even as an ideal. A dramatic fragmentation has occurred, and Canadians now divide on first principles. Where there was once *con*sensus, there is now *dis*sensus.

This became clear to me as I watched reaction to the "Freedom Convoy" that converged on Ottawa in February 2022 to protest against forced vaccination. In my eyes, the convoy clearly manifested a large and vibrant new public. Its vibrancy was reflected in the effort, and the risk to livelihood, that was involved in getting all those big rigs rolling toward our capital city in the dead of winter; its considerable size was shown in the crowds that gathered on bridges and in parking lots along the route to cheer this spontaneous cavalcade on. But, when the truckers and their supporters got to Ottawa, they were not treated as an emergent public with something important to say. Instead, they were treated as an invading army, and, finally, as a grave threat to national security. "These people," Prime Minister Justin Trudeau said, are "often racists" and "misogynists" who "don't believe in science" and who hold "unacceptable opinions."[12]

The CBC clearly concurred. Its nightly television newscast, *The National*, set the tone for its coverage, on the weekend the trucks arrived in Ottawa, by interviewing a trucker who was not even in Ottawa and who opposed the convoy, rather than talking to one of its participants. At no point, thereafter, did the CBC acknowledge the protest as a political phenomenon that deserved, both by its size and its argument, to be carefully examined and interrogated. Nor did the CBC recognize the protestors as an incipient *public* to which the *public* broadcaster owed, by that fact, a certain obligation. Instead, the demonstrators were viewed and discussed entirely as an unfortunate outcropping of *misinformation*, or as a problem of public safety. What this said to me was that the CBC, as the public broadcaster, now only converses with the publics of which it approves. It also said that the CBC doesn't recognize the growing polarization of opinion within the country as something which it has an obligation—a statutory obligation, in fact—to address with an even hand and an open mind.

Many other contemporary questions resemble the issue of vaccine mandates on which the Freedom Convoy disagreed with the government. The defining feature of these issues is that they divide people according to their basic commitments, or cultural stance, and not just on the basis of differing interpretations of some agreed set of facts. Some of these differences have been growing and establishing themselves ever since the various cultural revolutions of the 1960s began to take hold; some reflect the so-called filter bubbles now curated by social media algorithms. The point, in either case, is that *worlds* are colliding. Worldviews, in modern Western states like Canada, have become incommensurable—they no longer possess any common term, or denominator by which they can be related to one another.

The easy response to this collision, and the one unfolding all around us, is mutual vilification. The prime minister calls the protestors bigots, and the truckers, in turn, wave their ubiquitous, F🍁CK Trudeau signs. This is comforting to each party but does nothing to address the widening abyss between them. The group with FREEDOM on their blazons represented Canadian society as a contractual bond between free, self-determining individuals—they were, in short, classic liberals—while the majority, who claimed that failure to get vaccinated was a punishable anti-social act, stood for a view of society as something like an immune system: a single, collective, and mutually responsible being, acting under the sign of *life*. Neither of these views can be judged, in some simple sense, as right or wrong. They stand on different grounds and are conditioned throughout by the grounds on which they stand. Their only possible *modus vivendi* lies in curiosity, mutual respect, and a willingness, as Leonard Cohen once said, to "compare mythologies."[13] This spirit was not evident in Ottawa in the winter of 2022—on either side.

This dissensus, as I've called it, is something new. Canadians have always disagreed, sometimes violently, but formerly they disagreed within an overarching modern consensus. When public broadcasting was born in the 1930s, Prime Minister R. B. Bennett presented it as an instrument for "the diffusion of national thought and ideals."[14] The man who led the lobby for public control, Graham Spry, saw the CBC, even more grandly, as a means by which Canada would realize its "destiny."[15] Both men saw Canada as a society animated by common ideals and bound for a common destination. They saw their country as developing within a broad, transnational consensus, whose pillars were science and democracy, progress and growth. These were *certainties*—constellations of ideas and practices that people didn't have to think *about* because they constituted the very framework within which they were thinking at all. These certainties allowed Canadian society to be conceived, despite all differences, as a shared field of action. Differences could be reduced to a common denominator, goods ranked on a common scale, and preferences plotted on a common grid.

Within this consensus, people might disagree about the pace of progress, the degree of democracy, or the proper weighting of a given scientific statement, but they agreed on the framework within which such disagreements remained contained. Prime Minister Louis St. Laurent could plausibly describe the Cooperative Commonwealth Federation (CCF, predecessor of today's New Democratic Party) as no more than "Liberals in a hurry," and the *Progressive* Conservative Party could be seen, by the same token, as just Liberals in favour of slowing down. Libertarians could still talk to social democrats because they saw each other's preferred policies as different ways of reaching a common destination. There was a common scale on which differences could be weighed, and this allowed media to practice a rough kind of fairness or balance in allocating time and assigning significance to events. This was the context in which service to the audience—to the public—made sense.

Now, the common denominator has gone, and consequently, people and positions tend to fall apart into hostile camps. The one who thinks that abortion is contrary to natural law and divine ordinance has no common ground for negotiation with the one who thinks it is both a legal and a human right.

The libertarian and the social democrat no longer have a common destination in mind. Informed consent and vaccine mandates quickly come to blows.

The modern order has broken down and now faces, in whichever direction it looks, stark contradictions. "A world ends," says American poet Archibald MacLeish, "when its metaphor has died."[16] The images and convictions that held Western civilization together, and allowed the commensuration of differences within it, have lost their compelling vitality. Economic growth faces such stringent ecological limits that it can no longer serve as the foundation of social justice. Progress trembles at the boundary of domains like genetic engineering and artificial intelligence in which humanity itself is in question. Democracy is threatened by media technologies that overwhelm and disable deliberation. Science, expected by its founders to calm the war of opinion, now inflames it instead.

Looking closer to home, we find that the legitimacy of Canada itself is in doubt. More than once, in recent years, I have heard the phrase "so-called Canada" pass without comment on CBC Radio. Universities, streets, and schools are renamed; statues of Samuel de Champlain, Queen Victoria, and John A. Macdonald are pulled down or defensively put into storage. To most of its citizens, the history of the country has become a blank and indifferent slate, easily rewritten according to the politics of the moment. This too seems to bespeak a fundamental disorientation.

What this suggests to me is the need for a searching reconsideration of where we have been and where we are going. But this is not what the CBC has been offering. Instead, a mood of reaction has taken hold, and the CBC has joined with many others in trying to shore up the foundations of the failing order. In journalism, objectivity has been refurbished, and campaigns have been launched against misinformation, as if all disagreements turned on simple matters of fact. The contradiction between growth and conservation has been papered over with green technology and promises of sustainable development. Science has been restored to its pristine lustre, with everyone expected to blindly "follow"—even when the science in question is comprised of speculative computer models whose probative value lies just north of tea leaves or bird entrails.

When "a society finds itself in crisis," Mexican poet and essayist Octavio Paz once said, "it instinctively turns its eyes towards its origins and looks there for a sign."[17] That seems to be what's happening at the moment. Faced with the collapse of modern certainties, it is easier to retrench on the original vision than to seek a new path. Threatened orthodoxies revert to idealized versions of themselves that have been simplified and streamlined almost to the point of caricature. Journalists lament the golden days before disinformation—when truth was their polestar. Defenders of science bemoan the lost era when a dutiful public attended faithfully to what *science says*. Heresies proliferate, as orthodoxies narrow and positions polarize. The loss of an overarching framework that all share makes any non-approved difference threatening. This creates a paradox: a social order that sanctifies diversity and encourages the elaboration and embellishment of new identities grows at the same time increasingly narrow and intolerant of any dissent from its first principles.

Reaction of this kind is not an adequate response, nor a proper exercise of the CBC's mandate. Canada is threatened. We have suffered the disorienting loss of our history, which makes us, as a country without a usable past, a country without a usable future. We are facing growing political polarization, as is evident in the treatment of the generally moderate Freedom Convoy protestors as enemies of the state, thus potentially making them, sooner or later, real enemies of the state. And we are threatened above all by our intellectual helplessness in the face of the new forces—political, ecological, and technological—that are remaking our world. We are trying, in the words of the New Testament parable, to pour new wine into old wineskins, with the result the parable predicts—the skins keep bursting.[18] Our ways of speaking and thinking are not adequate to the circumstances they are called on to address. The old gods—the old certainties—cannot save us.

Populism, CBC-style, is one of the old gods. A singular audience, presumed to share the same interests, the same tastes, and the same habits of mind, no longer exists. And yet the CBC continues to appeal to this lost ideal, addressing its listeners as one happy family who agree on all important questions. The result is that populism has turned, at last, into a new elitism, although an elitism of quite a different cast than the stuffiness, artiness, and neglect of popular culture that populists denounced in the 1960s and 1970s. The new elitism is essentially service to a single shade of opinion. Call it progressive, hip, woke, or what you will. The point is that discourse is taking place within the confines of a single club and seems very often intended to enhance solidarity within that club rather than to raise potentially divisive questions. Basic issues of definition and orientation are taken off the table as a like-minded *we* confers about what is to be done. Issues are foreclosed rather than opened.

It is time, it seems to me, for a new era—a third era—as distinct from the populist era, as the populist era was from Eric Koch's first era. Successive broadcasting acts have required the CBC to *inform* and *enlighten*, as well as to entertain its audiences. Latterly, they have also instructed the corporation to contribute to "Canadian unity." This is currently the law of the land. But how, at the present hour, is Canada to be enlightened and brought together? I think the CBC's first task, as an instrument of public purpose, should be the reconstruction of the country's public sphere, or public forum, as I will prefer to call it.

My essay on how this new era can be brought about will be divided into two parts. I'll begin with an impressionistic history of the CBC, which will sketch the leading characteristics of the first era and then show, in more detail, how the populist era has unfolded. This history will ground my account of possible futures. From its earliest days, the CBC has had an insecure and uncertain place in Canadian life, and any scheme of renewal ought to begin by recognizing this tenuous hold on public affections and political loyalties. I could easily make a separate book of the newspaper and magazine stories, written throughout the corporation's nearly ninety-year history, that have routinely found the CBC to be "in crisis," "under attack," or "in search of its soul." The godfather of public broadcasting in Canada, Graham Spry, was already announcing the CBC's terminal "decline" in 1961.[19] In 1993, Wayne Skene, a former director of television in British Columbia, went so far as to offer a "requiem."[20] More recently—in their book *The End of the CBC?*—David Taras and Christopher Waddell have disinterred the corpse, but only in order to rebury it, should their advice not be heeded.[21] The Conservative Party has been actively hostile to the CBC since the 1940s, and its most recent "Policy Declaration" promises to "rationalize"—that is, eliminate—"any programming that overlaps or competes with private sector equivalents," as well as to "reduce … reliance upon government funding and subsidy."[22] (Canada is already nearly lowest among Western countries in its funding of public broadcasting—only the United States and New Zealand do less, with the most generous countries, like Norway, Switzerland, and Germany, providing five times as much funding as the CBC receives.[23])

These historic difficulties need to be borne in mind when contemplating the future of the CBC—not as a counsel of despair, but rather as an indication that the case for public broadcasting has yet to be decisively made in Canada, and now needs to be made afresh in radically new circumstances. At the moment, unhappily, the CBC seems to have largely forgotten its past and, accordingly, allows past experience no scope in the imagination of its future. It does little to make its rich archive known and available to the public; it rarely consults older programs in shaping new work; and it fails to ask how past incarnations of its mandate can inform new accounts of its purpose. The result is disorientation—the effect one would expect from amnesia.

Part Two of my book will explore the character the CBC might have in a new era. Many Canadians now believe the CBC to be redundant and hold that is unfairly subsidized to do things that private media do just as well. Rabble-rousing calls "to defund the CBC" reliably excite crowds at Conservative rallies.

But it is also the position of the Conservative Party, enunciated in its most recent Policy Declaration, that the CBC, insofar as it acts as a "a true public service broadcaster," has "an important place in the Canadian broadcasting system." With the Conservatives campaigning to form Canada's next government, and perhaps ready, as promised, to "defund the CBC," it is a good time to ask the question that the party does not address in its Policy Declaration: What exactly is a true public service broadcaster? My contention will be that it will have to become an organization that is concerned with the curation of Canada's public conversation. In order to carry out this function, it will have to become, first of all, a peacemaker, interested in finding common ground among currently warring opinions. It will have to become much more radically pluralist than it currently is, hospitable to all publics, rather than identified with its preferred public. And it will have to open itself to new ideas rather than trying to refresh old certainties, learning how to think, and to question, and to talk in new ways. Journalist Robert Fulford, speaking of the CBC's first era, has said that the CBC was then "very close to the centre of Canadian intellectual life."[24] This is the position I think it needs to resume, in order to become, as the Conservative policy statement says, "relevant to Canadians"; and in the second half of my book, I will spell out the hints I have given here as to how this can be done.

The roots of this book lie in my forty-odd years of work at CBC Radio, between 1971, when I began freelancing, and 2012, when I retired as a broadcaster and producer at *Ideas*. Much of what I will say here, both about what *has* happened—and what now, in my view, ought to happen—has grown out my observations and my own experiments in program-making during those years, and I will sometimes allude to these experiences as I go along. I offer thanks, before beginning, to my colleagues at *Ideas*, who accompanied and assisted me through much of this journey. I particularly thank Bernie Lucht, my executive producer, and good fortune, for nearly thirty years. Without his discernment and trust, the ideas expressed in this book might never have found a place to flower and grow. And, finally, I thank my wife, Jutta, who has faithfully read, edited, and counselled me throughout this book's long gestation.

A last note: I am aware that *broadcasting*, in a physical sense, is on its way out, as streaming increasingly replaces the distribution of signals through the air. I will continue to use the word, nonetheless, in order to maintain historical continuity and because the term, in a broader sense, refers to all forms of diffusion, distribution, and dissemination.

PART ONE

1 | The First Era: A Brief History

> The boys my age in Toronto were fascinated by the wireless, and we'd pretend trying to talk through a hose was wireless ... The idea of communication, the taste for communication was in my being in some way. It was something in the age, world-wide communication. It was a fascinating time to be a young man.
>
> —*Graham Spry*[1]

The first public radio broadcast took place on Christmas Eve in 1906. It originated from what one historian calls a "transmitting shack" with a four-hundred-foot antenna that had been built for Canadian inventor Reginald Aubrey Fessenden at Brant Rock, Massachusetts.[2] Fessenden played a recording of "Oh, Holy Night" and read the Christmas story. Radio operators on United Fruit Company banana boats lying off Boston, amazed, heard Fessenden's broadcast and wrote to tell him so.

At first, radio was the province of the amateur and the inventor. The first full-scale, commercial radio station didn't go on the air until 1920, when Montreal's CXA, later CFCF, broadcast a live concert. A glittering crowd gathered for the occasion in the Château Laurier in Ottawa. The Prime Minister, Sir Robert Borden, was there, as was his soon-to-be successor Mackenzie King. "Montreal sings to Ottawa by telephone without wires," reported the Montreal *Star* the following day.[3]

Networks developed slowly. The first program heard nationally was on July 1, 1927, Dominion Day, as it was then called. The occasion was Canada's Diamond Jubilee, and a network of stations established by the Canadian National Railways carried the festivities live from Parliament Hill. The broadcast began with a carillon recital, recorded by a technician who climbed out among the gargoyles around the Peace Tower clock in order to get a good pickup on the bells. Later Prime Minister Mackenzie King addressed the gathering.

One observer, who was present, remembered "the enormous impression" the occasion made on "the press ... the politicians and particularly on Mackenzie King."[4] Speaking, for the first time, to everyone in Canada within reach of a radio, he experienced the promise and—who knows?—perhaps also the peril of radio broadcasting.

A year later, King appointed a Royal Commission under the chairmanship of Sir John Aird, the president of the Bank of Commerce, to advise the government on what its broadcasting policy should be. Britain had already opted for a publicly owned system, and the United States for unrestricted free enterprise. Canada, so far, had improvised. What finally made the government move was a controversy over religious broadcasting. Stations operated by the Jehovah's Witnesses had been accused of scurrilous attacks on other religious groups, notably the Roman Catholics. The government responded by closing the offending stations down. The opposition protested that this was an arbitrary decision and called for a comprehensive inquiry. The government complied.

The Commission's Chairman, Sir John Aird, wasn't particularly interested in radio—he'd had one once, he confessed to an acquaintance, but "threw the damn thing out"—and he was predisposed, on ideological grounds, toward private ownership. But he began to change his mind, according to broadcasting historian Frank Peers, during a visit to NBC headquarters in New York.[5] There, Aird was shocked to discover that the American network saw Canada as just another part of its market, and this offended his sense of Canada as a distinctively British society. An interview with Lord Reith, the founding head of the BBC, also helped persuade him that broadcasting ought to be a public responsibility. The terse twenty-nine-page report that Aird and his fellow commissioners submitted in September of 1929 urged the government to establish a national network of publicly owned stations. Canadian listeners, the Commission concluded, were not getting enough Canadian radio, and never would without public ownership. It also argued that the public option would be necessary to preserve a certain tone in Canadian life. Fearing the vulgarity of private radio, the commissioners called for the exclusive use of announcers with "cultivated voices." (At the time, BBC announcers, unseen though they were, were required to wear formal dress for evening broadcasts.[6])

The Radio League

The Aird Report couldn't have been released at a worse time. Two months later, the stock market crashed. Within a year, the government changed, and Mackenzie King's Liberals gave way to R. B. Bennett's Conservatives. At that point, the Aird Report could easily have been shelved had it not been for the formation of an organization called the Radio League. It happened on October 6, 1930, after a meeting of the Canadian Institute for International Affairs held at the home of Alan Plaunt, the son of a wealthy Ottawa family. Plaunt had come to admire the BBC's style of public service broadcasting while studying at Oxford. Graham Spry, who was one of those present, recalled:

> At the end of [the meeting], we stayed to have a drink, three or four of us. Tom Moore, president of the Trades and Labour Congress of Canada, Margaret Southam, daughter of Wilson Southam, one of the two owners of Southam Press. There was three or four of us stayed around, and I suggested that [since] it was going to be a quiet winter in the middle of the Depression, perhaps we could take up the subject of broadcasting. We have the Aird Report, there's great public interest. The thing is to direct the public opinion to the Aird Report and make that an issue.[7]

Graham Spry was then the national secretary of the Association of Canadian Clubs, a national network of groups that fostered interest in public affairs and public service in Canada, a position that gave him connections across the country. Like his friend Plaunt, he had studied at Oxford, where he had belonged to a little expatriate society that also included Lester Pearson, Frank Underhill, F. R. Scott, Eugene Forsey, and King Gordon, all of whom would go on to play prominent parts in public life in Canada. This group was inspired by a budding Canadian nationalism that was beginning to sense new possibilities in the country. Shortly before he returned to Canada in the summer of 1926, Spry wrote in a letter to a friend: "This is a stirring age, and it is Canada's century … All who feel the spirit calling should take a share in the destiny that is Canada."[8]

The Radio League was something between a movement and a conspiracy—a "respectable intrigue," as Spry himself called it.[9] Spry and Plaunt were at its heart. Spry had family connections to the prime minister, and he took full advantage of them. In an interview I recorded with him a few years before he died, he told me that he kept track of Bennett's movements from a vantage point in the Rideau Club, across the street from the prime minister's office in the East Block of the Parliament buildings. In this way, he learned that the prime minister, once a week, took "some sort of a massage or hot treatment" at the Château Laurier. "We used to watch that," Spry recalled, "and get around into the Château ahead of him, watch what stairs he was going down, and then come up the stairs as he was going down. And if he was in any way receptive at all, why, we would put some point to him."

In this way, Bennett gradually came around to the League's point of view, swayed by many of the same considerations that had earlier changed Sir John Aird's mind. Both men instinctively preferred private enterprise, but fear of American influence and the damage it might do, both to the British connection and to the tone of Canadian life, finally predominated. This persuasion was not shared throughout the Conservative Party, and Bennett soon faced questions about the extent of support for public broadcasting in Western Canada. He called on Spry, interrupting him one night at dinner and asking him to produce evidence that such support existed. Spry responded immediately:

> I said, well, certainly, I will leave tonight. Those were the days of trains. I ... telegraphed to Premier Brownlee in Alberta, whom I knew quite well. I didn't know the premier of Saskatchewan, but I knew the attorney general through the Canadian Clubs. And of course, Premier Bracken in Manitoba. One of my first political activities was the United Farmers of Manitoba in their campaign which was to carry them into office.[10]

Spry's connections carried the day. He returned to Ottawa triumphant, with resolutions in support of public broadcasting from the legislatures of Manitoba and Alberta and the cabinet of Saskatchewan. The League was also able to gain the backing of farm organizations, the Trades and Labour Congress, church organizations, and others. Another crucial piece in this growing coalition was the newspaper publishers. They saw private radio as a competitor for advertising revenue, and Plaunt and Spry did "a masterful job," in the words of broadcasting historian Michael Nolan, of playing on this anxiety. It appeared that public broadcasting had broad political support. But Michael Nolan warns against the conclusion that Canada "got public broadcasting because of a ... groundswell of cultural feeling." He thinks it was because "established economic interests," and above all the newspapers were, for the time being, in favour.

On May 16, 1932, R. B. Bennett introduced a bill creating the Canada Radio Broadcasting Commission (CRBC). The country, he told the House, must be "assured of complete Canadian control of broadcasting, free from foreign interference or influence." Without such control, radio would never become "a great agency for the diffusion of national thought and ideals or the strengthening of national unity." But the reality did not match the rhetoric. What the Radio League had wanted was something like the BBC: an independent crown corporation responsible to its own board of directors. What it got in the CRBC was virtually a department of the government.

The commission was not well enough funded to establish the network of public stations that the act envisioned, and the private stations it had to depend on to carry its programs proved balky. According to historian Frank Peers,

> They stirred up the hostility of many of the private station operators when they tried to bring them into line … b[y] mak[ing] rules that uniformly applied across the country. The private stations had not been used to any kind of supervision of this kind and resented it hotly and complained, of course, to their local members of Parliament, who would raise all these complaints in the House of Commons. The single House of Commons committee that reviewed the CRBC's work in 1934 was very hostile.[11]

During the short, unhappy career of the CRBC, the Radio League went into hibernation. Plaunt and Spry bought a rural newspaper called the *Weekly Sun*, once the organ of the United Farmers of Ontario. Spry became active in the CCF, a new political party founded in the same year as the CRBC was created. (It renamed itself the New Democratic Party in 1961.) Plaunt got involved in the New Canada Movement, an agrarian youth movement that began in Grey and Bruce counties in the winter of 1933 and swept through the counties along the shores of Lake Huron and Lake Erie. The movement seeded folk schools and study groups in farming communities and, as one farm newspaper commented, "stir[red] young farmers to go on a crusade to awaken the youth of the land to the need of building a new Canada."[12]

The CRBC Becomes the CBC

A second chance to create the kind of public broadcasting the Radio League had hoped for came in 1935 when the CRBC unexpectedly blew itself up during the election campaign of that year. The commission had broadcast a series of programs produced on behalf of the Conservative Party, but not identified as such. They involved a cracker barrel philosopher called "Mr. Sage." In the course of the broadcasts, "Mr. Sage" made a number of derogatory remarks about the opposition leader, Mr. Mackenzie King. King was furious and "vowed," says Frank Peers, that "if he got back into power, the regulatory agency that had allowed this disgraceful exhibition on the air would be decapitated."[13]

Within weeks of the "Mr. Sage" broadcast, King was prime minister again. Alan Plaunt seized the day. Spry was now publicly associated with the explicitly socialist CCF and, consequently, persona non grata with the new government, so he laid low and worked behind the scenes, while Plaunt took stage centre and worked tirelessly among the Liberals. The relationship that Spry had had with Bennett in 1932, Plaunt now had with King.

Plaunt's patient work bore fruit on June 15, 1936, when a new Broadcasting Act was introduced into the House of Commons, creating the Canadian Broadcasting Corporation. The basis of the act was a detailed memorandum that Alan Plaunt had submitted to the government the year before. The actual bill was drafted by Brooke Claxton, an associate of King's who had also been a Radio League member. Claxton said later that it contained 90 percent of what the League had wanted. There was to be an independent board, assured funding, a strong general manager, and—scarcely believable today—regulatory control of private broadcasting. Alan Plaunt was appointed to the new board. Its chairman, Leonard Brockington, inaugurated the new service on November 2, 1936, with an extravagant flight of populist rhetoric. "We hope," he said, "that men in lonely places will be able to tell us of their adventures, that the man in the ditch, the western farmer on the prairie, the fisherman, the fur trader, rich and poor, great and small, Canadian men and Canadian women, will share with their fellow citizens the perennial marvel of the human interest of their struggles and their achievements."

The first order of business for the board was to establish a chain of high-powered regional transmitters, which involved a running battle with one of King's most powerful ministers, and the one responsible for the CBC, C. D. Howe. By 1938, the Quebec and Ontario stations had gone ahead, but Howe was stalling the approval of a 50,000-watt transmitter for the prairies. According to Graham Spry, the matter came to a head when Howe ordered the board to increase the power of a private station on the prairies:

> The board decided that they would arrange an interview, and they did, with Mr. Howe. Brockington phoned him, and it was arranged for 9 o'clock the next morning … The whole board went to C. D. Howe's office, and Mr. Howe was told by Brockington that they had considered their oath, considered the Act, they had taken legal advice, and with the greatest respect, they would advise the minister that they had not the power to obey his instructions and he had not the power to give them. And Howe said, well, I am the minister, you read the Act, I am the responsible minister, and I will have to stand by my instructions. And according to the legend—there's nothing in writing that I know of to support this, but it's the story of the time—Mr. Brockington then said "Well, Mr. Minister, that being the case, we will simply hand you the resignations which we

> brought with us and resign as a complete board." Again, according to the story, C. D. Howe, being the man he was, threw up his hands and in a very agreeable smiling manner, said, "Okay, Brock, you win."[14]

A year later, CBK went on the air from Watrous, Saskatchewan, with an announcer speaking proudly of the "new modern transmitter building," from which he was broadcasting—"a new landmark in the centre of Canada's famed wheat growing country," with its "great vertical antenna reach[ing] 460 feet into the sky."

Meanwhile, those programming the new network were debating an issue that still sounds familiar today: How much American material to carry? Ernie Bushnell, the network program director, recalled discussing the matter "very thoroughly with … the whole board of governors" and reaching a decision to "interlard" the new Canadian shows with popular American offerings. "Sure it was probably the American shows that attracted the audience in the first place," said Bushnell, "[but] if you put a good show back to back with an American show, you had an audience." Many people transferred their allegiances in this way, according to Bushnell, "and it built up the audience tremendously."

At first, these programs were produced by very few people. James Finlay, who had worked for the CRBC in Montreal, was a CBC producer in Vancouver:

> [I was] responsible for all of the music, all of the drama and all of the talks and everything else that went [on]. There was one other chap, Mercer MacLeod, who was at the station, and between the two of us, we produced all the programs—all of the programs. As a … producer, it was my job to set up the chairs for the orchestra, put up the microphones, and during rehearsals, waltz the microphones around until I achieved a good pickup. I finished up in 1940, and I was producing three half-hour dramas and eleven musical shows per week. That was my production. That's why I worked 90 hours a week.
>
> There was a sense of give and take and camaraderie that existed throughout the corporation. Mind you, it was a relatively small group. There were only 200 people on staff from coast to coast. We were all caught up with the same single purpose of … building what became a very proud organization.[15]

The greatest event in the early years of the CBC was the Royal Tour of 1939. It was the first visit of a reigning monarch to Canada, and with the undivided attention of the country, the corporation went all out. Three years before, 100 people had been nearly the entire staff of the CBC. Now it needed that many just for the tour, a marathon of six gruelling weeks, 7,000 miles, and ninety-one broadcasts. Practically, every step their majesties took was theatrically narrated. The opening broadcast—of their disembarkation at Quebec City—gives the flavour:

> Evidently the King and the Queen are just about to come out the side of the ship ... The officer is saluting at the side of the vessel, just where the gangway opens into the body of the ship. The photographers are poised here. The Prime Minister and [Ernest] Lapointe remove their cocked hats. Here comes the King, down the gangway, followed by the Queen. The King saluting the Prime Minister. His foot is on Canadian soil and the Royal Tour has begun. There goes the first gun of the salute from Citadel Hill ...[16]

The Second World War, which followed only months after the royal visit, reinforced the central position the Royal Tour had given the CBC in Canadian life. With more than a million Canadian boys and men in arms over the next five and a half years, listeners hung on Lorne Greene's nightly news bulletins and on Peter Stursberg and Matthew Halton's dispatches from the front.

But the war also brought bitter conflict within the corporation over the issue of the independence of the CBC. The CBC's first general manager, Gladstone Murray, a British Columbian and a legendary flyer during the First World War, favoured close cooperation with the government and its war aims. Alan Plaunt, the dominant figure on the board, was a neutralist. He had been instrumental in hiring Murray as the corporation's first general manager, luring him back to Canada from the BBC, where Murray was director of public relations. Plaunt had initially admired Murray's good program sense, as well as his gift for public relations, but now Plaunt took issue with his former ally. Along with other intellectuals of the time, like Frank Underhill and Frank Scott—both part of his and Spry's Oxford circle—Plaunt feared that war might tear Canada apart again, as the conscription crisis had done during the First World War. He also felt strongly that all points of view should continue to be aired on the CBC—access for representatives of the CCF during wartime was a particular bone of contention—and he disliked the change of tone he detected in CBC broadcasts, as war approached. Early in 1939, he wrote to Murray about a special New Year's program the CBC had just presented. "As a combination of banality, bad taste, cheap sentimentality and jingoism," he said, "this program would be hard to beat. An American listener would get the impression that we were a bunch of mawkish and sentimental imperialist halfwits with no independence or identity of our own."

Murray was just as adamant. With war imminent, he visited England to confer with Lord Perth, the head of propaganda for the British government. He connected with Canadian spymaster Sir William Stevenson, now famous by the code name "Intrepid," and he advised the government on questions of people's loyalty. When the outbreak of war brought censorship to the CBC, he shrugged off its impact. This is evident in a network broadcast he made in 1941:

> While the demands of war are paramount, there is no prospect that the fulfillment of these requirements will damage the worthwhile distinctive tradition which Canadian broadcasting was evolving before this war. Now, propaganda as a term is always suspect, and rightly so, because it is the stock in trade of Goebbels' lie factory. As I see it, our task is simply to tell the truth, to make it luminous and attractive, and to repeat it, and to go on repeating it, until the lies and vain boastings of the enemy have dissolved as the morning mist before the penetrating and wholesome rays of the rising sun.[17]

Plaunt didn't share Murray's view that war had suddenly turned the world black and white, and conflict between them continued. But Murray, for the time being, held all the high cards. Being an instrument of "national thought and ideals," as Bennett had promised in 1932, changed its aspect in wartime. Plaunt resigned from the board. A year later, he died of cancer, after several unsuccessful operations, still a young man, he and his wife expecting a baby in just a few months. His board colleague Leonard Brockington, paying tribute, said that Plaunt died with "his best ideas unrealized but not uncommunicated."[18]

Propaganda was certainly a big part of the CBC's job during the war. It was propaganda for a good cause and perhaps just as "luminous and attractive" as Murray predicted, but it was still unabashed and uncritical promotion and lionization of Canada's war. My old *Ideas* colleague, Lister Sinclair, then just starting out at the CBC, became one of the "Achtung Brothers"—in demand, along with his colleague Jules Upton, for his ability to perform a passable and appropriately sinister German accent. He worked on shows like *Fighting Navy* and *L for Lanky*, in which actor Neil Leroy gave voice to a Lancaster bomber. The producer of *Fighting Navy* was so punctilious about the correct naval atmosphere, Sinclair recalled, that when a taxi was called, his assistant was required to say, with a straight face, "Duty boat alongside, sir."

But, at the same time, the CBC was growing up and becoming a much more confident organization than it had been when the war began. In 1940, when Alan Plaunt resigned, Gladstone Murray had the full support of his minister, C. D. Howe, and the rest of the board. Two years later, Plaunt was posthumously vindicated. Led by a number of Plaunt's old allies, the House of Commons Broadcasting Committee chastised Murray and reaffirmed the independence of the CBC.

The CBC Wins Its Political Independence

Heartened, the CBC began to assert its right to present critical opinion, and this led to a series of well-publicized battles with the government. The first to become public was in the summer of 1943. The issue was a series of discussion programs that were being planned by the talks department and were to be called *Citizen's Forum*. Writer Morley Callaghan had been hired as moderator; Ernie Bushnell was then the program director; and Neil Morrison was the supervisor of talks. Forty years later, in an interview with me in the old CBC Studios in the Château Laurier, Morrison recalled how the row began:

> Bushnell said to me, look, you and Morley had better go up to Ottawa and see if you can line up some people, cabinet ministers and parliamentarians and others, for the series. We had a list of topics. So, Morley and I got on the train and came up to Ottawa, and got a room in this hotel, the Château, and we started out. We decided, as is proper in terms of protocol, to start with the government, and we tried to start with Brooke Claxton, who was then parliamentary secretary to Mackenzie King. I knew him personally, and he was busy, couldn't see us until the afternoon. So instead of wasting the whole day, we said alright, let's go and see the Conservatives. So, we called, I think it was Ross Brown in the Conservative office, and we went to see [Conservative House leader] Gordon Graydon, and we got an appointment with [CCF leader] M.J. Coldwell and with [CCF National Secretary] David Lewis. And we talked to them, and tentatively we lined up some names. And in the afternoon, we saw Claxton, who had an office in the East Block. We came in, we explained what it was all about and what we wanted, and we showed him, or Morley, I guess, showed him a list [with] Coldwell, and others on it. Claxton just hit the ceiling, hit the roof. This was an insult, we'd come to them last. The thing went from bad to worse. Morley got annoyed and lost his temper, and he got up and went over to the desk in his belligerent, pugilistic style and leaned over and said, "Mr. Claxton, maybe you can tell

> Morrison what to do. He works for the CBC, but I'm a taxpayer and you can't tell me what to do." And I was saying to Morley, "Morley, sh! sh! sh! Take it easy, Morley, take it easy." Then Claxton got mad, and he was mad already, so the fat was in the fire, and we didn't succeed in convincing him of our good intentions or integrity.[19]

Morrison found the story funny in retrospect—and it is—but, at the time, it was an important turning point in the life of the CBC. Claxton wasn't just unhappy to have been consulted last, he was unhappy about CCF views being aired on the CBC at all, and he followed through by getting in touch with General Laflèche, the minister of National Defence, the department then responsible for the CBC. Laflèche, in turn, put pressure on the corporation's acting general manager, Augustin Frigon, formerly of the Aird Commission, and planning for the program stalled. Then the Winnipeg *Free Press*, not by accident, learned of the affair. Other newspapers picked it up. Frigon denied that there had ever been political interference, and the series went ahead.

This was a partial victory only. The following year General Laflèche tried again. This time, the issue was a Montreal reporter's criticism of the management of New Brunswick mental hospitals. Neil Morrison was once again involved:

> Ken Johnson, who was working for the [St. John] *Standard*, was doing a broadcast for us, which was very critical about the Liberal administration in New Brunswick. And the premier, I think it was, called Laflèche, and Laflèche called the studio in Montreal and got through to the control room and told them to cancel the broadcast—which was a flagrant interference which he had no right at all to do. That got out in the newspapers, and for the second time, Laflèche was castigated in the press. It wasn't by accident it got out, although it would have been hard to prevent it, because after all, Ken Johnson was a reporter and I wasn't going to let anything of that sort pass, either. So Laflèche didn't survive very long, and that was one of the last times a cabinet minister made an attempt, because, politically, you got your fingers burned.[20]

Another wartime incident involved the Minister of Justice and future Prime Minister, Louis St. Laurent. After a disturbance at the Stoney Mountain penitentiary in Manitoba, he asked that the CBC no longer cover such occurrences. The acting general manager, Augustin Frigon, was happy to comply, though the policy was never applied. Nevertheless, it was changed in 1946, when former newspaper Editor Davidson Dunton became the chairman of the CBC board of governors:

> I came into the chairmanship, I must say firmly committed to the independence of the CBC and that it would also be visible, and found that the management had agreed to a request from the then minister of justice that news of prison riots not be broadcast on the CBC. This was because in other penitentiaries, the CBC was the only radio they were allowed to listen to, and it was thought that if they heard of violence in one prison, that would tend to incite or lead to violence in another, and in effect, that request was being met. I found this out and brought it up at the board, and we just stopped the practice right away and so advised the minister of justice.[21]

Dunton remained chairman for the next twelve years, and from time to time, he still had to fend off political interference. On one occasion, the prime minister himself wrote to complain about a certain commentator and asked that he be taken off the air. The letter "found its way into the hands of the press," Dunton recalled, and Mr. St. Laurent had a "rather uncomfortable afternoon" in the House. In another case, an attempt was made to cancel a documentary on unemployment in Cape Breton:

> C. D. Howe was acting prime minister and called me and said, I hear you're doing this thing on unemployment. It's not to go on. I said, Mr. Howe, it's the board of governors that finally decides things. And he said, okay, you'd better resign. That was the end of it. But to his credit, he called me back the next day and said, you know, I'm sorry, that was the wrong thing to do, and we ran the program. But I think that was a good reminder, when C. D. Howe backs off.[22]

Davidson Dunton certainly deserves the credit he gives himself here for keeping C. D. Howe and other cabinet ministers at bay and instituting what came to be called the arm's length relationship between the CBC and the government. But the battle would have to be fought again after the Conservatives regained power in 1957. In 1959, *Preview Commentary*, a daily broadcast of political analysis and opinion, was cancelled on direct government orders, and it took the resignation of the supervisor of Talks and Public Affairs, Frank Peers, and his entire staff to get the show reinstated.

The cancellation of *Preview Commentary* was the last instance of direct government meddling in programming decisions. This did not mean, however, that the CBC stopped policing itself, in the interests of a cordial relationship with the government, as an incident that occurred in 1955 showed.[23] In that year, Bernard Ostry and Henry Ferns published a controversial book called *The Age of Mackenzie King*—a work that was critical, and even somewhat sarcastic, about the former Liberal prime minister whose astonishing longevity in office—twenty-two years in all—had made him a sacrosanct figure in his own party. The CBC organized a panel discussion about the book. One of the people the producers invited to participate was Brooke Claxton, a consummate Liberal insider, a former cabinet minister, and an intimate of King's. Claxton was incensed that the CBC would even consider discussing the book. He exploded, upbraiding the producer who phoned him so loudly that he hoped, he said later, that "everyone at the CBC could hear him." He also wrote, in the wake of this call, that he "could not imagine Dave passing on this." Dave was Davidson Dunton, a friend, a Rockcliffe neighbour, and a member of the same club. Later that day, the two friends ran into each other. Dunton said he hadn't heard of the intended program but would look into it. The next day, when the project "came across his desk," he cancelled the broadcast.

This is as much as the record shows. Did Dunton peruse the book in the meantime and reach his own judgment, as journalist Grant Dexter, a firm Liberal, claimed? Or did he take Claxton's word that it was a disreputable book, better ignored? I imagine he assessed the merits of a book he may himself have considered scurrilous, weighed them against the hostility the broadcast might stir up within the fading but still formidable Liberal political hegemony in Ottawa, and found discretion to be the better part of valour. Bernard Ostry was not slow to inform the Conservative opposition, who framed the cancellation, in the House of Commons, as another instance of the CBC's Liberal bias. The Tory press also took up the cry. The Saskatoon *Star Phoenix*, for example, commented that when the CBC cancels "two carefully prepared and scheduled programs which were to deal with a supposedly controversial biography of the late Prime Minister Mackenzie King" and then gives "exceedingly woolly and confusing" answers to questions about the matter, the public could "hardly fail to suspect the worst."

The affair interests me because it shows so clearly the limits of the independence that Dunton achieved and successfully defended. A quiet word at the club was still possible, even after overt orders from ministers of the government had been discredited. The government's disposition still mattered and still had to be carefully discerned by the CBC's executives. The CBC might safely stand off the Social Credit member of the House Broadcasting Committee who told Dunton in 1951 that a broadcast with Montreal psychiatrist Ewan Cameron savoured unacceptably of communism and atheism, but it continued to operate within a consensus with evolving but still definite boundaries. In 1955, Davidson Dunton judged *The Age of Mackenzie King*, with its harsh attack on the received view of King as a friend of the people, to be still beyond the pale.

The CBC Puts Down Roots

During the period of Dunton's chairmanship, the CBC grew steadily in confidence and scope. One of the ways in which it built its influence and its independence was by putting down roots in rural Canada. This connection to the country began in earnest with the inauguration of farm broadcasting. The man responsible was Orville Shugg. Shugg had been a farmer at Watford, Ontario, and a friend of Alan Plaunt's, whom he had met through the New Canada Movement in the early 1930s. Through his friendship with Plaunt, Shugg began to think about what public broadcasting could do for farmers, and when the CBC was set up in 1936, he communicated his ideas to the general manager, Gladstone Murray. Murray encouraged him, but it was two more years before he actually let him go ahead. The first broadcast went to air in 1939.

One of the distinctive features of the CBC's farm broadcasts, the precursors of today's Radio Noon, was their use of dramatic serials. These were modelled on the popular dramatic serials of the time—the "soaps," as they came to be called, because they were so frequently sponsored by soap manufacturers—and they provided virtually the first steady work Canadian actors ever had. The stories revolved around fictional farm families in each region of Canada, and they constituted, Orville Shugg told me years later, "propaganda in the best sense of the word. Propaganda for better farming practices, better farm life, better country life, better public participation, and more public participation of farmers and farm families in the affairs of the community, all that sort of thing." These broadcasts had an impact that's hard to recapture in today's wired world, Shugg said:

> To get a proper view of what the situation was then, realize that of the Canadian population in 1936, 46 percent were rural. That's hard for people today to conceive of. Also, it's difficult to grasp the isolation that existed in Canada between regions and even between rural communities. That's why radio was so important when it burst on the scene with a service specifically designed for farm people. I was absolutely amazed at what happened in the Maritimes, for instance, where the isolation was

> even greater than it was in the outback of Ontario, if I may call it that. In the Maritimes, the farmers had been at the mercy of itinerant drovers [who bought their] livestock. The itinerant drover[s were] really preying on the farmer[s] because they had the market prices, the farmer[s] didn't. As soon as we came on the air with daily market prices, the itinerant drover could no longer drive into a farmer's yard and say I'll give you such-and-such for such-and-such at such-and-such a price. The farmer'd say, oh no, you don't, I just heard on the CBC air that the price should be such-and-such. We, in fact, we, the CBC Farm Broadcast put the itinerant drovers off the road in the Maritimes during the first year of operation, and of course we were heartily blessed for that by the farm people of the region.[24]

The next step in the CBC's outreach to farmers was *Farm Radio Forum*. It was set up in cooperation with the Canadian Association for Adult Education and went on the air early in the war. Discussion groups of up to twelve people were formed across the country, and they chose the topics of the broadcasts. Each week a pamphlet, outlining the issue, was sent out in advance of the program. The CBC, according to Alan Thomas, one of the producers, "knocked itself dead to get ranges of opinion." Following the broadcast, the groups would confer and then submit their conclusion to the program. Once a month, a summary of these conclusions was broadcast. After a discussion of what Canada was fighting for in the war, for example, the program reported that 25 percent of participants had said that it was just as important for Canadians to defend themselves against "a repetition of the Depression" as it was to defend themselves against fascism.

According to Alan Thomas, later the head of the Canadian Association for Adult Education, *Farm Radio Forum* had so many "spin-offs … that it would be impossible to count them all." People organized for discussion remained organized for other purposes, and the sense of recognition that went with being consulted helped to shape the political atmosphere of post-war Canada. Thomas gives the example of the "hundreds of arenas" that got built at the initiative of what had begun as *Farm Forum* groups—arenas that became community meeting places as much as hockey rinks. Later when *Citizen's Forum* was created, and groups formed in cities and towns as well, these repercussions expanded. A *Citizen's Forum* group in Toronto, Thomas remembered, became the seed of Canada's first public housing association, and presently its first public housing project.

Citizen's Forum also organized and broadcast public meetings. Frank Peers, as a producer in the Talks and Public Affairs Department, often moderated and introduced the speakers at such meetings. "I would say," Peers recalled, "that our chief thrust was to establish and maintain connections with various segments of Canadian society … We collected opinions that were expressed by organized groups, but we also took our broadcasts out to public meetings [where we worked] with business and labour groups, [did] programs in the field of human relations [and] women's interests … and so on … [W]e had very close ties with a whole range of organizations … that were otherwise not concerned with broadcasting, but which we thought should have a part."

Harry Boyle joined the Farm Department, from a background in private radio and newspapers, in 1942, remaining at the CBC until he was appointed to the Canadian Radio and Television Commission (CRTC) in 1968. He remembered the CBC at this period as permeated by what he called "a strong degree of social consciousness." "The way I'd describe it," he said, "is that there was a kind of consciousness in all of the people involved in these programs that something had been wrong with the cycle in the country, that people had suffered too much in the Depression, that we mustn't forget about it during wartime, and that we must be sure in the post-war situation that we don't go back to the bad old thing. And you heard that expressed. It was there." This sense, Boyle said, that a recurrence of "the bad old thing" must be prevented, and that the CBC had a leading part to play in preventing it, led to "an enthusiasm for the whole concept of broadcasting." This was expressed not just in *Citizen's Forum*, with its explicitly political subjects, but in many programs. Elizabeth Long, who was put in charge of "Women's Talks" at a time when no such talks existed on the CBC, developed, according to her colleague Marjorie McEnaney, "a wonderful coterie of women broadcasters," whom she encouraged and supported. This led in 1952 to the introduction of *Trans-Canada Matinee*, the first daily magazine-style broadcast on the CBC. Helen Carscallen, who got her first job at the CBC on *Trans-Canada Matinee*, remembered it as a show that enlarged "women's window on the world" and created a community of interest amongst women "who were isolated."

Canada, before radio, had been a country of regions—not just two solitudes, but five or ten, depending on how you count. The CBC became its first national cultural institution. "The CBC was the theatre," says Neil Morrison. "It was the opera house, it was the music hall, and this is what I think the CBC did in those days, was to hold an image up to the country in a whole variety of ways, and people said yes, this is here, this is where we are, this is what we are. A lot of people had a sense of mission about what they were doing in terms of the life of the country." "The CBC became a kind of mecca," Harry Boyle says, "back in the late 40s and early 50s for people who wanted to sell a story, if you wanted to get … a creative job … if you wanted to write music … it was a better bet to get a piece of music commissioned at the CBC than anyplace else in the country." Drama prospered extraordinarily with more than three hundred plays, many original, presented on CBC Radio in the 1947–1948 season.[25] Christopher Plummer, who took part as a young actor, recently recalled this period in radio drama as "Canada's most creative time in the performing arts"—a time when we found our "own voice" and our "own identity." He did not mean, he emphasized, that Canada had not had many notable artistic accomplishments in the meantime, but only that this was "the closest thing Canada had ever had to its own identity, artistically speaking." "Nothing has happened since," he added, "to carry that on."[26]

Intellectual culture also flourished. "In the 1950s," according to journalist, broadcaster, and critic Robert Fulford, "the Talks and Public Affairs department of CBC Radio … seemed to me to be very close to the intellectual centre of Canadian life." Series like *Architects of Modern Thought* introduced Canadians to major contemporary thinkers. It was through this series, Fulford recalled, that "George Grant, who was then a young professor at Dalhousie, first came into my consciousness in the mid-50s when I sat down and I heard a half-hour lecture he gave on the CBC on Jean-Paul Sartre. And then I dug out other lectures of his, and … I've followed him ever since." (This remark was recorded in 1986, when Grant was still alive.) Grant, who had worked on *Citizen's Forum* in the 1940s, was particularly closely connected to the CBC—two of his books began life as CBC series, *Philosophy in the Mass Age* as a set of lectures for *University of the Air* in 1959 and *Time as History* as the Massey Lectures ten years later—but many other Canadian thinkers were also first introduced to the broader Canadian public through the CBC and then heard regularly there—Northrop Frye and Marshall McLuhan are notable examples. Fulford continued:

> At no time did I think it even slightly remarkable that I was listening to a young philosopher from Halifax explain Sartre to me on the radio in a half-hour lecture. That seemed exactly what the CBC wanted to do and was meant to do. And then on Sunday afternoon, for example—at 4:30 on Sunday afternoon on *Critically Speaking*—you would hear three critics, one or two of them first class, who would address the new books, the new films, the new music being written in Canada, the new broadcasting programs, television and radio programs. So it seemed natural if a critic of some interest was emerging somewhere in Canada [that you would hear them on the CBC]—I remember, for example, Chester Duncan from Winnipeg, who I thought then, and think now, was one of the most interesting commentators around. Chester Duncan was regularly on

> CBC Radio. That was my only exposure to him for 10 or 15 years. I only knew him as a voice from Winnipeg. So, I can't say that the most important thoughts of Canada were conceived in the CBC, but the CBC seemed to be the place where the Canadian thinkers, academic and otherwise, were expressing themselves to a large public, or a fairly large public.[27]

None of this should be taken as saying that CBC Radio, at this period, was carrying all before it. Enlarging the bounds of public discussion provoked resistance and controversy. Lister Sinclair's play *Hilda Morgan* is an example. Originally written for radio, it was adapted for television in 1952, CBC Television's first year of operation. The play touched, very circumspectly, on the subject of abortion. After its broadcast, George Drew, the leader of the Conservative Party, then in opposition, rose in the House of Commons to denounce Sinclair's "disgraceful play" and called on the government to take action to prevent "filth of that kind going out over the televisions of this country." This was not untypical. The CBC had a liberalizing influence on Canadian public opinion in the 1950s, but it was still often attacked in the House of Commons and the press for its godless, freethinking ways.

The drawing power of American popular entertainment also continued to be an issue. The CBC pioneered its own variety shows, like *The Happy Gang*, and its own "soaps," like *John and Judy*, but still carried many popular American offerings like *The Carnation Contented Hour*—the name came from the sponsor's boast that its milk came from "contented cows"—and *Lux Radio Theater*. It also depended on the commercial revenues these shows, foreign and domestic, generated. A line-up clogged by these sacrosanct and regularly scheduled shows, which had to be "programmed around," Harry Boyle remembered, made it difficult to find room for weightier material. This difficulty led, in 1947, to the introduction of a commercial-free evening into the schedule: *CBC Wednesday Night*. Benjamin Britten's opera *Peter Grimes* had its North American premiere on *Wednesday Night*; T. S. Eliot's verse play *Murder in the Cathedral* was presented; and live classical concerts of both orchestral and chamber music were offered. Harry Boyle, the show's first producer, recalled that the program often found listeners who hadn't previously known that they liked that sort of thing. "I've had countless cases," he said, "of people telling me that the CBC annoyed them with the symphony or … recital … or so and so, and then [they] gradually found themselves listening and involved."

The CBC was creating a distinctive place for itself in Canadian life, and it did so by distinguishing itself both from its commercial milieu, where popularity was the *sine qua non*, and from its political milieu, where partisanship was the norm. CBC broadcasters saw themselves as public servants who were exercising a calling rather than as advocates or as entertainers. Bernard Trotter, who joined the talks department in the early 1950s, conceived his department's role along the lines of an intellectual and cultural utility:

> I wasn't an ideologue of any kind, but I did very much believe in a national press in Canada ... we had to connect with each other and we had to share experiences and ideas and talk to each other in various parts of the country, and the CBC seemed to me to be a very good medium for this ... I thought that the CBC was a pipeline, a communications system. I didn't have any notion of trying to shape what people thought in any direct sense. I thought that our role was to make sure that the important subjects were explored and discussed and that all valid—well valid's a loaded word—that all points of view were given a chance to compete with each other in a free forum. I suppose, in a sense, the Athenian *agora* was the ... theoretical model. Perhaps I was too naïve about this, but I really did think of myself as a neutral or a neuter as far as imposing any of my own ideas on the country. That wasn't the way I saw my role at all.

The CBC as Regulator

Another important aspect of the CBC's existence during its first era was its role as a regulator. It is now little known and little remembered that the CBC was created in 1936 as the orchestrator and organizer of the entire broadcasting system. According to the Broadcasting Act of that year, private radio stations were to serve only as local and subsidiary elements within a system that the CBC would coordinate and direct in the public interest. The CBC had "complete authority over the formation and operation of networks" and the power, subject to appeal, to "prescribe the periods to be reserved by private stations for CBC programs."[28] Radio frequencies were considered a public resource, which could not be owned as of right but only used subject to certain revocable terms. This was the theory. What happened in practice was often dramatically different.

From 1936 onward, both the influence and affluence of the private stations steadily grew, and, as they did, the CBC's regulatory powers became more and more difficult to exercise. In 1948, for example, the CBC tried to assert its right to take over the class A broadcast channels then occupied by CFRB Toronto, CRY Winnipeg, and CFCN Calgary. These were called clear channels because they offered a greater reach and better clarity of reception than the AM frequencies then occupied by the CBC. In a broadcast system in which the public interest was paramount, and the CBC was accepted as the legal and proper guardian of this interest, there would have been no question that the CBC should have access to these superior channels. What happened in fact was that the CBC was forced to retreat in the face overwhelming opposition from the newspapers, the Conservative Party, and the Canadian Association of Broadcasters (CAB), the voice of the private station owners. (By this time, the newspapers no longer feared private radio as a potential competitor, as they had at first, and, after 1945, they tended to react more as fellow capitalists when the private stations' interests were seen to be threatened by burdensome CBC regulation.) So complete was the CBC's defeat on this occasion that it not only gave up any claim to these channels, it also granted the private stations new powers. The main one was that CFRB was allowed to increase its transmitter capacity to 50,000 watts—a concession that soon had to be extended to other stations and one that altered the balance of forces within the broadcasting system at a stroke, beginning the trend from public to private predominance that continues to this day.[29]

During the years between the end of the Second World War and the eventual removal of regulatory authority from the CBC in 1958, the CAB lobbied relentlessly against the pre-eminence the 1936 Broadcasting Act had assigned to the CBC. The slogan and theme of this prolonged campaign was that the CBC could not be both "cop and competitor." The argument made sense if you accepted its premise—that the private stations and the CBC possessed equal standing and equal right and should therefore "compete" on a "level playing field." According to this premise, the frequencies the private broadcasters occupied on the electromagnetic spectrum were "resources," which they were entitled to appropriate, own, and enjoy like any other form of property. The CBC had its piece, they had theirs, and may the better man win. But this was not the premise of the 1936 Broadcasting Act, as I noted earlier. Its presumption was, first, that the air is a common possession that can be used only as a conditional privilege not as a right, and, second, that broadcasting must be considered an integrated undertaking in the public interest, with the CBC at its head and private stations in subsidiary roles under its governance.

In 1932, when public broadcasting began, R. B. Bennett spoke grandly of "national … ideals" and "complete Canadian control" and then introduced the weak, under-financed, politically hamstrung CRBC. Graham Spry, whose patient courtship of Bennett had led to this contradictory result, later spoke delicately of a "conflict in [Bennett's] soul." This conflict was not just in Bennett's soul but in Canada's. Public enterprises, like the CBC, have shaped Canada—from the days of canal building in the early nineteenth century to the creation of Canadian National Railways, Trans-Canada Airways, and the various provincial Hydros in the twentieth century. Our geography, our relatively sparse population and smaller market, and our proximity to the American dynamo have all urged these public undertakings. When Graham Spry addressed the parliamentary committee on broadcasting in 1932, he told its members that the choice before them was simple. "The question is," he said, "the state or the United States?" The same choice has been posed, again and again, in many other spheres of Canada's economy and culture. But at the same time our predominant political philosophy, liberalism, with its preference for free markets and limited government, has argued just as urgently against state involvement. This has created a political split personality epitomized in the quasi-mythical figure of the Red Tory.[30] Canadian playwright and political commentator Herschel Hardin, an admirer of Canada's distinctive tradition of public enterprise, has gone so far as to call liberalism "the American ideology in Canada" because, under the peculiar conditions of Canada's national existence, free markets have almost always equalled American dominance and control.[31]

This conflict in Canada's soul was vividly played out in the 1940s and 1950s. At the same time that CBC Radio, and, after 1952, CBC Television, were giving Canada a new sense of itself as a country, the owners of private stations were fighting aggressively to contain, limit, and reduce the CBC's role in the broadcasting system. They were aided and abetted by the Conservative Party, which was sympathetic to these aims on both philosophical and practical political grounds. Philosophically, the Conservatives were predisposed to free enterprise. Politically, they associated the CBC with the long Liberal hegemony under which they had chafed from 1921 until 1957, with only the relatively brief interruption of the Bennett government between 1930 and 1935. Now and then, this political enmity found real occasions—a CBC decision in 1943 to cancel the broadcast of a speech by Conservative leader John Bracken as "too political" created lasting resentment, as did the cancellation of the scheduled broadcast on Ferns and Ostry's *The Age of Mackenzie King*.[32] But, for the most part, the Conservative Party's ill will was free-floating and persistent, no matter how scrupulously the CBC maintained its political neutrality.

From 1944 onward, the Conservative Party argued that the CBC should be stripped of regulatory power and an independent regulatory authority created. The argument was always that the CBC could not be, at once, "cop and competitor, judge and litigant," as future Conservative leader John Diefenbaker told the House Broadcasting Committee in 1944.[33] The opportunity to make this change came after the 1958 election. The Conservatives had already formed a minority government the year before, and, when the

Liberals unwisely triggered a second election, John Diefenbaker's government was returned in the largest landslide in Canadian history, taking 208 seats to the Liberals' forty-eight. From this dominant position, it was easy to enact the long-promised reform of broadcasting, and the new government was quick to do so.

The Broadcasting Act of 1958 created a new body called the Board of Broadcast Governors (BBG) and invested it with the regulatory powers that had formerly belonged to the CBC. This not only freed the big private stations from the tutelage of the CBC, as they had long and vocally desired, it also subjected the CBC to this new authority. Liberal leader Lester Pearson, in a lengthy and thoughtful speech to the House of Commons during debate on the bill, presciently identified the weaknesses of this new set-up. First, he described the fateful philosophical change that was being enacted. In the legislation of 1932 and 1936, broadcasting had been considered a privilege that must be exercised in the public interest. In the intervening years, he said, it "had gradually become a vested interest and eventually ... a right." Pearson then predicted the rancorous relationship that has existed ever since between the CBC and the regulatory authority nominally placed over it—first the BBG and, after further modifications in 1968, the CRTC (at first the Canadian Radio and Television Commission, and after 1976, when its mandate was expanded, the Canadian Radio-television and Telecommunications Commission). The difficulty, he went on, was that the CBC had its own legislative mandate and its own board, as well as its own reporting line to the government, so it was unlikely to submit to another board telling it what to do. What was being created behind the fiction of unity embodied in the BBG, Pearson said, was, in fact, "a dual system." In practice, he concluded, the new board would "tend to become a regulatory body for private stations only, influenced increasingly by the financial situation of those private stations."[34] To put it very simply, capital had been put back in the driver's seat, and the CBC turned into an aberration within the system it had once epitomized and defined.

This, of course, had not happened all at once. Looked at cynically, the Broadcasting Act of 1958 can be seen as only signifying and ratifying what had been the case from the beginning. The CRBC was less than a year old when hostile members of R. B. Bennett's government tried to deprive the fledgling commission of half the monies appropriated for it, while the prime minister was away in England.[35] The idea of the CBC as a regional service supplemented by local private service was erased at a stroke in 1948 with the decision to allow CFRB to increase its power to 50,000 watts, the maximum permissible. Graham Spry, looking back ruefully near the end of his life concluded: "The CBC has never been created as provided by law, and its powers have been progressively or regressively reduced, or in some way impeded, their use impeded. The concept of a single national broadcasting system with an almost unlimited number of low-power, local stations in all sorts of ownership forms, cooperatives, municipal, private, advertising, what you will—This, of course, never came into being." What ultimately stunted the CBC, Spry said, was "the power of Canadian wealth"—"more millionaires were made out of radio broadcasting than ... out of building the C.P.R." (This was said, obviously, with reference to a time when "millionaire" still signified great wealth and not just ownership of a house in Toronto or Vancouver.)

Perhaps the CBC never had a chance and came, as Spry intimates, still-born from the womb of the 1936 Broadcasting Act in which Alan Plaunt had invested so many of their mutual hopes. Still, much was accomplished in the years before 1958. Davidson Dunton, the chairman of the CBC board between 1945 and 1958 is hardly an impartial witness, but he believed that Canada had created a unique hybrid: "I don't think any other country in the world ... achieved as good a combination of sponsored and non-sponsored programs, commercial and non-commercial programming, and of cooperation [between] privately owned stations and publicly owned stations. And I think just taking that period of radio, Canada, for its size, for its complexity, for its two languages, had a darn good radio system. It compared with anything in the world." This is the other side of the coin—both in relation to Spry's pessimistic assessment that the CBC was "never created" and the perfervid rhetoric of the CAB, in which the CBC usually appeared as a brutal bureaucratic juggernaut. The smaller private stations, in fact, often talked out of both sides of their mouths—complaining about the CBC on the one hand, and benefiting from their affiliate status on the other. "In lots of ways," Dunton said, "they were partners in the same job. We wanted them to do a good job, and they wanted the CBC to do a good job."

The BBG Gets to Work

This cooperation did not entirely end in 1958—the CBC still counted many affiliates as part of its radio and television networks—but a direction was set toward a predominantly private system in which the CBC had little chance to set the tone. Virtually the first order of business for the BBG was to begin hearing applications for "second" television licences—i.e., licences for private stations in cities already served by the CBC. From the time television began in Canada in 1952, the government had been unwilling to fully fund the expensive new service. Some powerful ministers in the St. Laurent government, like

C. D. Howe, had felt that television should be left entirely to private capital, but a strong recommendation from the Massey Royal Commission had convinced the government that the CBC should remain in charge. The result was that the mixed system that Dunton praised in radio was initially established in television, with CBC owned and operated stations in the biggest cities and private affiliates in smaller centres. The stations that the BBG began licencing were of a new kind—independent private stations that would soon assemble themselves into the rival CTV network, founded in 1961. All the new licencees made glowing promises about the new Canadian programs they would produce and then quickly reneged, filling their prime-time schedules instead with popular American shows—a pattern that has held with virtually every private television station licence granted since.

As Pearson predicted, the new regulator became responsible for the private broadcasters only. The CBC, as I noted earlier, received its money and its mandate directly from Parliament and was not disposed to be bossed around by this inexperienced newcomer. The new private stations, on the other hand, were the Board of Broadcast Governors' direct creations, and in practice, the board tended to become protector and preceptor to them alone. The board was composed of three full-time members and nine part-time members. Among the part-time members were Guy Hudon, the dean of law from Laval, and Eugene Forsey, then research director for the Canadian Labour Congress. Both soon noted the growing closeness between the board and those it regulated, as Eugene Forsey remembered:

> I think there came to be a good deal too much pally-ness [i.e., familiarity] between the private broadcasters and some members of the board. This irked Dr. Hudon and myself particularly. On one occasion, we had a particular broadcaster before us, not in a public hearing, and everybody was addressing him by his first name. Everybody except Dr. Hudon and myself at all events. And Dr. Hudon said in his most pointed manner, "Mr. So-and-so, if I may be permitted so to address you." And I know we were both getting increasingly uneasy about what seemed to us the pally-ness which existed between some members of the board and the people who came before us. A regulatory board is set up to regulate, and I think that should be made abundantly clear to people who are appointed to it, that they are expected to stay at arm's length from the people they're regulating.[36]

The creation of the BBG introduced ambiguities into the broadcasting system that still exist today. On top of its own board, the CBC now had an additional board between itself and the government. Before, when the CBC had wanted to extend its service, it had asked the government. Now it had to ask the BBG, and sometimes to fight with the BBG, even to provide basic coverage. For example, the CBC had no station in Quebec City, and so, in 1962, as CBC President Alphonse Ouimet remembered, it found itself before the BBG pleading for the right to establish such a station:

> We had one place [in Quebec] where we had a station and studios—Montreal. We had five places for English-speaking Canada ... I had to appear before the BBG on three long days as a witness presenting the case for the need of the national service to have a station in the capital of the province of Quebec. And our application was being opposed by a small local private station with good connections. And it was crazy. There was the CBC saying we need a station here because we must have more French originations and not all from Montreal. There is talent here to develop, and so on. While somebody else was saying boy, if you let me come here ... I'll make so much money.... And we barely got it. Two of the governors had to resign.[37]

One of the governors who resigned was Eugene Forsey:

> The thing that I resigned over was the question of a French television licence for the CBC in Quebec City. This came before us over and over and over and over again. I can't remember now how often. And we were given to understand that there were all sorts of complications about general principles of broadcasting which had to be worked out, and there might be this and that and the other development in the future, and we couldn't really say. And finally, Guy Hudon and I lost all patience, and we didn't think that the reasons given for delay were valid and we were tired of being told next time, there'll be a decision, next time, there'll be a decision, next time, there'll be a decision. But we could get nowhere. Our colleagues simply did not agree with us and we thought the time had come for a decision, we thought the decision ought to go to the CBC. So, we got out.[38]

Forsey and Hudon's drastic action did eventually lead to the CBC getting its station in Quebec City. But, with the establishment of the BBG, and the consolidation of the fledglings it had licenced and nurtured into the CTV network three years later, a new broadcasting environment was created for the CBC. The 1936 Broadcasting Act was informed by the principles that had inspired and guided Graham Spry, Alan Plaunt, and their allies in the Radio League: that the air is a commons, that all private appropriation of it must subserve the common good, and that this public purpose must be determined by a public agency with the authority to realize it. These principles were formulated at a propitious moment. Radio was an exciting novelty, still able to induce that "taste for communication" that Graham Spry felt to be part of his "being." Capitalism was still haunted by its catastrophic failure in 1929. And the nationalism that stirred thoughts of destiny, "the spirit calling," and a New Canada in Plaunt and Spry was still able to make common cause with the waning Empire nationalism of men like Bennett and Aird. By 1958, broadcasting was a business, the BBG were on a first name with the representatives of the "industry," and the CBC had become the sore thumb which it has been ever since.

This anomalous position was expressed in two ways. The first was that a single system had become a dual system, without this dual, and divided character ever being acknowledged. Was the BBG, later the CRTC, actually in charge? Not really. First of all, it quickly fell victim to the well-described process of "regulatory capture" by which the regulator's main duty becomes the economic welfare of those it ostensibly regulates. Indeed, the BBG was more or less born captured since the Conservative government, in a sense, set it up to lend the appearance of public purpose to the granting of private television licences. And, second, the CBC, as I've stressed, had its own legislative mandate, its own reporting line, and its own hard-won independence—all of which made it adamantly and sometimes fiercely resistant to being governed by its new "governors." But the CBC was no longer in charge either, as I've said. Quite the opposite. The many private television stations and networks that now began to come into existence all became, in effect, American broadcasters in Canada. And this meant that popular taste, to which the CBC also had to cater, was more and more shaped by American programming. So, the CBC was progressively less free and less able to lead. The fiction of a single "broadcasting system" was maintained as a political shelter for the government of the day, but, since neither the CBC nor the BBG/CRTC had any real capacity to prescribe its objects or set its standards, its integrity as a system was really nothing more than a sentimental relic. The full consequences would take many years to emerge, but the Radio League's dream would endure only as a carefully preserved façade.

2 | "We Got to Have Them Watching": From the Beginning of *Television to the End of This Hour Has Seven Days*

Television broadcasting began in Canada on September 6, 1952, when the Minister of National Revenue J. J. McCann inaugurated station CBFT in Montreal. Two nights later CBLT went on the air in Toronto—with the station identification slide displayed backwards and upside down. With well-established American border stations already beaming a full service into Canada, the CBC was behind from the start, and sometimes the haste showed.

One faction in the government, led by the influential C. D. Howe, would have preferred that the CBC stay out of television altogether. Howe's opinion was that it was too expensive. "If private operators think it worthwhile to risk that much money, let them go ahead," he said, but, if he were living in his hometown of Port Arthur, he continued, "I'd kick like a steer at paying taxes to bring television to Montreal and Toronto."[1] Howe's opinion was outweighed by the view of the Massey Royal Commission on National Development in the Arts, Letters, and Sciences, which had reported the year before. Its chairman, Vincent Massey, once a member of the Radio League, was a strong supporter of public broadcasting. The commission convinced a majority in the government with its argument that private television would purvey mainly American programs that would not serve "national needs."[2] The CBC was put in charge of the development of television.

But, as had been the case in radio, government support remained stingy and qualified. The CBC was always what Harry Boyle once called "a skim milk operation." TV was initially financed by a 15 percent excise text on the sale of new televisions, but no permanent provision for predictable and reliable funding was ever made. So, when the initial boom in television sales inevitably subsided, the CBC was left to rely on annual appropriations—an insecure, hand-to-mouth existence that has continued to this day. CBC Television, like radio before it, was forced to rely on advertising to make up the difference between the cost of serving so vast and so various a country and the inadequate public revenue provided for the task.

The "mixed system" that had evolved in radio was reproduced in television. "Owned and operated" stations were created in major centres, while a patchwork of affiliates served the rest of the country. A more or less continuous negotiation took place between the CBC and its affiliates and advertisers over where the balance between popular programs and more elevated fare was to be set. Opinion has divided over whether this fostered a typically Canadian genius for accommodation and compromise, or just produced an attenuated system forever falling "between two stools"—never serious enough for some, never popular enough for others. What is certain is that by 1958, when the Conservative government introduced its new Broadcasting Act, the CBC had lost whatever power it once had to influence where this boundary between popularity and public purpose was to be set. The creation of private networks, purveying almost nothing but popular American programming in prime time, exerted strong pressure on the CBC to conform to the American standard. The CBC had never had much ability to use what BBC founding director Lord Reith once called "the brute force of monopoly" to educate public taste. After 1958, with New York or Los Angeles now just a flick of the wrist away, it lost this capacity altogether.

But TV in Canada, as I've said, was behind from the beginning. By the time the first two CBC stations began broadcasting for three hours a day in 1952, the American networks were already providing Canadians within reach of their signals with a full twelve-hour-a-day service and had been doing so, in some cases, for several years. The effect of this prior formation of taste was even more pronounced than it had been in radio. TV production was a lot more expensive and elaborate than radio production, and the higher gloss the American networks could afford was more noticeable. Invidious comparison was inevitable.

CBC Television flourished, nevertheless, and those who were there often recalled its first years as a time of great adventure. Broadcasts frequently went live to air; fine arts consorted with popular and folk arts. One night might feature *Tosca* or *Hamlet*; the next, *Country Hoedown* and *Don Messer's Jubilee*. An average year saw about two hundred new productions, almost all of them one-offs rather than parts of series.[3] Drama was a particular strength, with offerings ranging from beloved classics by Molière and Ibsen, to new plays by Canadian authors, to the work of celebrated contemporary dramatists like Jean Anouilh, Arthur Miller, and Luigi Pirandello. British communications scholar Graham Murdock, writing in 1980, claimed that, in drama, the CBC outperformed the BBC in both originality of concept and execution.[4] The BBC bought Canadian productions in those years, and CBC Television's supervisor of drama, Sydney Newman, was eventually enticed to England, first to Granada and then to the BBC.

The rapid development of television also created a cultural collision within the CBC, as the new TV people were thrown together with the old radio people. J. Alphonse Ouimet felt the impact in his bones. He had been with the CBC from the beginning, joining its predecessor the CRBC in 1934 and becoming the corporation's general manager in 1953 and its president in 1957. He was also a distinguished electrical engineer who had helped design, build, and demonstrate Canada's first television in 1932 and coordinated the CBC's planning for the new medium between 1948 and 1952. He compared the early years of television to having "a tiger by the tail." The American head-start, and the novelty of the new technology, imposed haste and improvisation from the outset. "All we could do," Ouimet recalled, "was to try to get the beast to go in the right direction." New people, with film or theatre experience, had to be brought in fast, and "we were not able to give the amount of introduction to the CBC that we would have wished." The public service ethos in radio had amounted, in Ouimet's recollection, to a religion—he spoke to me about that "old … religious approach that existed in radio"—but the new TV people, in his estimation, "were interested really in one thing, and that was to make television programs. They didn't care whether it was CBC or any other institution."

Journalist Robert Fulford spent a lot of time at the CBC in the 1950s—observing and taking part in programs—and he agrees with Ouimet. According to Fulford, the run-in between television and radio provides a key to understanding "a good deal of what has happened to mass culture in our time, or to the distribution of ideas in our time." He continues …

> [The] collision [between] all these adult educators and … upper middle brows who staffed the CBC Radio service and these new people in television … was really traumatic for all concerned. On the one hand, the radio people felt invaded, and I can remember a button being worn around the CBC in the early 1950s saying "Help stamp out TV." … For the television people who were conquering this brave new world—and who knew that television is not a medium like the others, it has its own logic, its own rules, its own beauty—for these television people to be governed by these old men of 40, 45, these old men who knew nothing but radio

> and adult education and in any case were veterans of the Second World War and looked back on a shaping radio experience ... was terrible ... On the one hand, the old adult educators, the informers and the idealists, if you like. On the other hand, the busy opportunists who wanted big-time careers in television and wanted to graduate from Toronto or Vancouver or perhaps Montreal, to New York or Los Angeles or London, and in many cases did so. The collision between these two forces was traumatic for all of them.[5]

This cultural clash was intensified by the mystique of the new medium with its high costs, its technical demands, and its seductive promise of power and glory. Lister Sinclair, who had written and performed in radio since his debut as one of the Achtung brothers during the war, remembered being told that he was too old to qualify for the television training course that was being organized by Mavor Moore and Stuart Griffiths in the run-up to CBLT's launch in 1952. Sinclair was thirty-one at the time, but it had been resolved, he says, that nobody over the age of thirty was to be admitted to the course because television was a young person's game.

The CBC Centralizes

Another difficulty that the CBC faced in the 1950s stemmed from the decision to centralize the corporation's management in Ottawa. Until 1953, the CBC networks had been run from their broadcasting hubs. The chairman of the board, Davidson Dunton, was in Ottawa, the nominal headquarters, but Augustin Frigon, the CBC's general manager during the ten years before his death in 1952, had established himself and much of his administration in Montreal, and the English network was run from Toronto. This changed with the appointment of Alphonse Ouimet in 1953. Dunton offered Ouimet the choice of remaining in Montreal like his predecessor, but Ouimet replied, as he told me later, "I am not going to be a general manager in Montreal. I'm general manager for the whole thing. The head office is here, I'm moving here."

This was a logical decision, and Alphonse Ouimet was, by all accounts, a logical man—logical to a fault, said his detractors. It made sense that if Canada was to be a unified country, and the CBC a unified corporation, that the headquarters should be in the capital, and not in one or the other of its non-communicating "solitudes." But Canada has never been a logical country, and Ouimet faced opposition from the start. "I was not successful," he remembered ruefully. The idea that French and English broadcasting should be under the same management was unpopular. Senior staff refused to move. Richard Lambert, the supervisor of School Broadcasts, complained of French autocracy and led a little revolt in Toronto. A series of articles in *Le Devoir* made the opposite reproach: that rule from Ottawa signified English control of the French network.

One of the people who did agree to go to Ottawa with Ouimet was Neil Morrison, the supervisor of Talks and Public Affairs. He was French-speaking, he had lived and studied in Montreal, and he shared Ouimet's hope of building bonds between French and English. It was his view that broadcasting, so far, had "strengthened the two solitudes" by reinforcing existing Quebec nationalism and planting the seeds of an incipient English Canadian nationalism. He tried to manage both networks together and to introduce common fare, but "it just didn't work," he said.

Management from Ottawa not only failed to bridge the national divide, it also created new problems. The executives tended to lose touch with their programmers, while increasing their contact with the politicians who were now their neighbours. "You couldn't walk down the street from the head office in the Victoria Building to the Château [Laurier] cafeteria," Morrison recalled, "without running into senior civil servants, or members of the government, or the opposition, and you constantly got criticism." The worst of this, Morrison thought, was the loss of perspective: "In Ottawa, ministers and deputy ministers are much more important than they are anywhere else in the country, so you have an atmosphere which makes it difficult to maintain the autonomy and independence of a broadcasting system."

The disadvantages of being in Ottawa were forcibly brought home to the corporation's senior managers on December 29, 1958, when seventy-four Montreal television producers walked off the job. Television had been a wild success in Quebec where *Radio-Canada*, the CBC's French service, had the field to itself. *Téléromans* (TV novels), as they were called, brought the province to a virtual standstill. Fernand Quirion, who directed *La Famille Plouffe*, one of the first and most popular of these series, recalled that, during the Monday evening broadcast, "stores would close and … you couldn't get service in a gas station because people were watching television." But the producers were unhappy at how they were treated by *Radio-Canada* management in Montreal, which was headed by Alphonse Ouimet's brother André.

Their strike was different from the beginning. It was, as was later said, "a gentlemen's strike," with many of its protagonists already rich and famous.[6] Actor Jean-Louis Roux, then a household name in Quebec as Ovide Plouffe, recalled the amazement of a policeman assigned to the strike when a producer pulled up to the picket line in a fancy red sports car and began serving the strikers cognac from a built-in bar. "Christ," Roux overheard the cop say, "what are they wishing to have? Do they want the building itself, or what?" And the strike quickly acquired political significance. Grievances about workload and how contracts were negotiated soon turned into questions about union recognition and about Ottawa's supposed indifference and incomprehension. The striking producers gained and held the support of colleagues at *Radio-Canada*, including prominent journalists like Gérard Pelletier and René Lévesque, and won the backing of a who's who of the emerging Quebec intelligentsia, notably Jean Marchand, Pierre Trudeau, and André Laurendeau. Lévesque, in an interview many years later with CBC Radio's *Morningside*, attributed the beginnings of his nationalism to the strike:

> ... one thing ... I can't forget was going to Ottawa with a few hundred of the guys during the strike and knowing damn well that the whole of the French network was going to pieces, and finding such complete indifference. That was Diefenbaker's government at that time, and the guy we met was Michael Starr. He was minister of labour. But anyway, it was like a man from Mars as far as we were concerned. He couldn't care less. He had nice words, like politicians will have when they want to get rid of you, but there was no understanding of what was going on and the implications. So, I got mad.[7]

Lévesque was, as he wrote in *Le Devoir* at the time, "permanently disgusted." Roger Lemelin, the author of *La Famille Plouffe*, writing in the same paper, construed the "serene ... indifference" with which he claimed the federal government and CBC management had reacted to the strike as a sure sign that both wished to see "the means of expression of French culture weakened." André Laurendeau was similarly disposed to a nationalist interpretation of the strike. Within a year, René Lévesque was a minister in the new Liberal government of Jean Lesage and had given the "quiet revolution" its not-so-quiet slogan: "*maîtres chez nous*," masters in our own house.

The strike was settled in early March of 1960 with a partial victory for the producers. They won the right to collective bargaining but bowed to management's adamant demand that they not affiliate with the *Confédération des Travailleurs Catholiques du Canada* (CTCC), once known rather derisively as the "*Syndicate des Curés*," the priests' union, but now perceived by the CBC as both radical and political after fighting the famous strikes at Asbestos in 1949 and Murdochville in 1957. Both sides agreed that things at *Radio-Canada* were never the same. Friction between producers and management continued; the chasm widened between the French and English networks; and the camaraderie that had been part of life *Radio-Canada* in the 1950s—"the old spirit," in Alphonse Ouimet's words—never returned. Jean-Louis Roux had supported the strike, but he agreed with Ouimet in at least this respect: "I … didn't find back … the same enthusiasm and the same involvement that existed before the strike. There wasn't the same feeling as a team as we used to have … and I don't think that feeling was ever, ever found again."

The End of the First Era

By common consent, the producer's strike ended what Eric Koch calls "the first era" at *Radio-Canada*—the era characterized by a sense of mission and shared commitment—the "feeling as a team" of which Jean-Louis Roux speaks and which Alphonse Ouimet even dared to call that "old religious approach." In Toronto, the first era would end in a different way, with the battle that began in 1964 between Ottawa management and the producers of the weekly television newsmagazine *This Hour Has Seven Days*. In the producer's strike, unionization and the divided loyalty it would potentially introduce were the great issues. In the case of *Seven Days*, what was at stake was "objectivity"—the appearance of impartiality that Ottawa management insisted that the CBC must maintain and that *Seven Days* executive producer Douglas Leiterman claimed was no more than a "myth," a protective façade concealing a profound allegiance to the status quo.

The story of *Seven Days* goes back to the beginning of current affairs television. As had been the case in radio, which relied entirely on Canadian Press for its "news bulletins" during its first years, independent journalism developed relatively slowly at CBC Television.[8] One of its pioneers was Ross McLean, who produced Canada's first national newsmagazine, *Close-Up*. Among its young producers, when the show debuted in 1958, were Patrick Watson and Douglas Leiterman. Watson had joined the CBC in 1956 as "a young intellectual fresh out of graduate school" and was initially impressed, as he recalled in his autobiography, by "the tremendous intellectual buzz of the place."[9] Leiterman came from a newspaper background and had first begun to appear on the CBC as the national correspondent for the Southam News Service. Working together under McLean's mentorship, they soon discovered a common bond: both believed that television could change the world.

By the time the two men met, Watson had already come to the conclusion that "TV was not the medium I had expected it to be … [It] was not primarily a vehicle for informing people (not, that is, in the somewhat academic and arrogantly teacherly way I had expected to use it) but was something closer to theatre."[10] This insight, he says, began in disappointment and then gradually developed into "a theory and a strategy."

> In the beginning, I had this almost evangelical notion about … television … thinking what a tremendous medium to spread enlightenment amongst the masses. And I can recall becoming terribly disillusioned within the first year, during which I was producing and directing a lot of panel shows, discussion shows, really interesting people getting together to discuss really interesting issues. You remember, it was the year of Suez, it was the year of Hungary, it was an incredibly interesting time. And I would come out of the studio after putting one of these things live on the air, just so stimulated with the discussion, only to find out that nobody I talked to could remember anything anybody had said. And it was particularly frustrating because, if the program was very reasoned and people were being very careful and thoughtful and generous with each other in the discussion, and intellectually very fascinating, I found that nobody had watched at all. But if people were shouting and yelling at each other, then people watched and told me that it was a great program and they really liked what's his name, you know, the one on the left, but they couldn't remember anything that was said, although perhaps they could remember that the subject was Suez—or was it Hungary? So that led me to rethinking what the nature of the information was that we were purveying …[11]

Watson's colleague and interlocutor Douglas Leiterman had come to a similar insight by a rather different route …

> My notion was that there must be some way to translate journalistic skills into a visual medium which would have a lot more impact, and that really was what I was interested in. In fact, I might spend a year writing major front-page stories out of Ottawa every day, and in that time, I might do three pieces on CBC television, and I would go back to Vancouver and meet people who would say, "Gee, I saw you on television the other night, and I'd wondered where you'd been for the last ten years." Well, I'd been on the front page of their daily newspaper for the last ten years with a byline in about 14-point type, but they'd never noticed it. So, for someone who was interested in exploring the possibilities of change in our society, here was a magnificent discovery: that you could get people who were cab drivers, checkout clerks, to get involved in national political stories because of the merits of the visual medium, the sort of thing that they wouldn't read in a hundred years in a newspaper.

The conversation between Leiterman and Watson that began at *Close-Up* continued for several years. Leiterman, in the meanwhile, directed a number of celebrated documentaries. Watson produced a program called *Inquiry*, out of Ottawa. Their chance to produce the kind of current affairs television they'd been talking about came with the appointment of Reeves Haggan to the position of general supervisor of Talks and Public Affairs in 1963. Haggan was a man who shared Leiterman's hope that "cab drivers [and] checkout clerks" could be drawn to the CBC through television. According to Haggan, the CBC at that point talked only to what it called *its* "audience"—a conversation he described as "Don Mills talking to Don Mills"—the new Toronto suburb which stood, in Haggen's mind, for a complacent and somewhat ingrown sensibility. What many of his "colleagues and predecessors" had not understood, he said, was "that it is a dream to think that you can communicate with the public at large entirely in intellectual terms ... if you want to talk to the wider audience ... you have to first of all attract their attention, which you can do by amusing them, delighting them, frightening them, irritating them. Now, if you're prepared to do that, then you can convey an awful lot to them in the way of giving them a better understanding of what is actually happening, and enabling them, perhaps, to make better informed judgments on current affairs."[12]

Haggan's desire to unsettle what he saw as the cozy faculty club atmosphere of the public affairs department accorded perfectly with Watson and Leiterman's aims, so when they presented their proposal for a Sunday night newsmagazine, he quickly approved it. *This Hour Has Seven Days* went on the air in October 1964 with announcer Warren Davis reading the bold "manifesto" that Watson and Leiterman had written for the occasion:

> *This Hour Has Seven Days*, a show ranging over the complete spectrum of responsible journalism, of such natural interest, such vitality and urgency that it will recapture public excitement in public affairs television and become mandatory viewing for a large segment of the nation. *This Hour Has Seven Days*, probing dishonesty and hypocrisy, drawing attention to public wrongs and encouraging remedial action, presenting tough encounters with prominent guests hot in the news and prepared to be grilled.[13]

It was a little pretentious—"complete spectrum," "mandatory viewing," etc.—but it gave a fair taste of what was to come: politicians *were* grilled, wrongs *were* exposed, and the whole electrifying spectacle *did* create such public excitement that the show's final broadcast attracted three million viewers in a country, then, of less than twenty million.

Leiterman, Watson, and Haggan all agreed that the show must have what Watson called a "visceral" impact. The manifesto stipulated that this effect ought to be pursued within the bounds of "responsible" journalism, but "interest … vitality and urgency," Watson says, were still to be given the highest priority. A ruthless editing style imparted the desired kinetic energy …

> We trained ourselves to notice points at which our attention wandered off and we'd be thinking about … what we were going to have for lunch, or anything other than what was right on the screen, and make a note. So we developed a kind of house rule where if it couldn't hold the attention of the producers in the cutting room, it didn't go on the air—however relevant and important it might be in terms of the Argument, using the word in the eighteenth-century sense with a capital A …. We wanted to make television that just had people saying hunh! like that. And we would sacrifice a lot of stuff, and we made mistakes doing that. We left stuff out without which perhaps you couldn't understand the story. We said, well, finally it doesn't matter because if we don't do it our way, they're not going to watch, and *we got to have them watching*. They can ask the questions later.[14]

What mattered was what Watson called "visual coherence." An item must, as he said, "play." New recruits were "train[ed] … to lose their need to tell the whole story." "The important thing," Watson said, "was everybody was talking about it" and, to this end, "tremendous discipline" was exerted. Douglas Leiterman—"the guy who held us to that discipline," according to Watson—said more or less the same …

> We really tried to put on an hour of television where every minute of it would have energy, and where the viewer sitting at home would at least once or twice in that hour say "My God, how did they manage that?" Or, "That's what I think. How come nobody's ever said it before?" Or say to himself, "Boy, they sure got that guy where it hurts." Or "That's something I always wondered about." Or "Isn't it great that there are really people in the world like that person that I just saw on that program?"[15]

Heroes and villains were an important part of this formula. Among the heroes was the sewer cleaner who returned the $18,000 he found in the course of his job one day, and the Indian agent (sic) who bilked his own bureaucracy to build houses on a Yukon reserve. American consumer advocate Ralph Nader made his network television debut on *Seven Days* at a time when he was still shut out of the big American media following the publication of his exposé of the auto industry, *Unsafe at Any Speed*. Among the villains was the American Air Force pilot who was shown in Beryl Fox's remarkable Vietnam War documentary *The Mills of the Gods*. In a shocking scene shot live by cameraman Erik Durschmied from the co-pilot's seat during a bombing raid, the pilot obligingly provided his own unaffectedly triumphant commentary:

> Real fine, real fine, real fine. That was an outstanding target. Alright, we bombed first of all, and we could see the people running everywhere. It was fantastic! It's very, very seldom that we see Victor Charlie run like that. [Victor Charlie was American military slang for Viet Cong.] When they do, we know we got him. If we could keep him on the run, why we'd know we're gonna really hose him down. By jove, that's great fun. I really like to do that.

For the politicians, there was "the hot seat." It drew on a device that Watson had invented earlier for *Inquiry*, the political affairs show he had directed in Ottawa. A "periscope camera," as he called it, was placed out of sight in a "donut hole" cut in the table used for the show's panel discussions and then raised unobtrusively for "low-angle extreme close-ups." "I would hold back the periscope shot until the journalists were really boring in and the guest was feeling the challenge, sometimes sweating it, sometimes glorying in his or her deftness in fielding every … ball and then, relishing the exact timing of the cut, I would go to that big, tight, low angle, cinematic shot of the authority figure, dancing his dance or polishing her phrases."[16] Watson's contrivance was controversial in Ottawa, and, when he returned from his summer vacation after the show's first season, he found that "management had ordered the table [which housed the camera] broken up." The "hot seat," the form in which *Seven Days* revived the idea, was an isolated and exposed chair, brilliantly lit, with a rifle microphone pointed at it, where politicians were invited to show their good faith by submitting to interrogation-style interviews. In order to heighten the drama, Watson remembered, they "turned off the air-conditioning and got all the smokers (a majority in those days) into the studio just before a hot seat interview."[17] In the absence of a smoke machine, which they couldn't afford, the cigarette haze filtered the studio lights and provided the chiaroscuro, film noir effect they were looking for. Politicians soon learned to avoid *Seven Days*, and, for cabinet ministers, recalled Roy Faibish, who worked for the show in Ottawa and fielded the rejections, the first season wasn't even that old before that became an order.

Seven Days broke, definitively, with a carefully constructed CBC tradition of impartiality. Davidson Dunton had worked hard at the "arm's length relationship"—the entente whereby the CBC exchanged political neutrality for non-interference by the government. His successor Alphonse Ouimet had spent the years of the Diefenbaker government (1957–1963) trying to convince skittish Conservatives that the CBC was not a tool of the Liberal party—"After a government has been the opposition for 22 years like the Conservatives had been," Ouimet said, "they were convinced that everybody in the CBC were Liberal activists." Many people in the Talks and Current Affairs department shared the view I quoted earlier from producer Bernard Trotter: that the CBC should function, as he put it, "like a pipeline," which conveyed information and opinion to people as impartially as possible and allowed them to make up their own minds. So, not only were the young *Seven Days* producers a group of "cocky, unmanageable" brats, as Watson admitted ten years later in an interview with Michael Enright on *This Country in the Morning*, not only was sophisticated, *happening* Toronto pitted against stodgy Ottawa, and the TV vanguard against the old radio rearguard, but something utterly fundamental was at stake: Was the CBC only a conduit for opinion—each carefully weighed, marked, and balanced—or was the CBC itself a shaper of opinion? [18]

Management Draws the Line

Three weeks after the show went on the air, management staked out its position. Eugene Hallman, the vice president of programs, issued a statement of the CBC policy that he judged had already been threatened by the conduct of *Seven Days*: "No program," he said, "can be permitted to adopt an editorial point-of-view or take a position in matters of public controversy. CBC journalism is free to gather and present the facts of controversy in a systematic and unambiguous form, [but] comment on the issues under discussion by program staff is not permitted by policy."[19] The executive producer of *Seven Days*, Douglas Leiterman took a dramatically different view. Writing in the *Ottawa Citizen*, he claimed that objectivity and neutrality in journalism have "never been more than myths," which hide the reality of practical journalistic decision. "The protection of the public" can, therefore, only consist in a good "choice of the man who uses the tools," as well as a measure of trust in "the public's shrewd and certain awareness when it is being hoaxed."[20] Both positions, when simply stated, were highly idealized and somewhat naïve—Leiterman's confidence in his own righteousness and the public's keen nose, as much as Hallman's assurance that CBC producers could "unambiguously" and "systematically" assemble facts without in any way influencing them. But Alphonse Ouimet and his senior management group in Ottawa never budged during the two short years of the *Seven Days* "war," as it was often called. Ouimet told the parliamentary committee that looked into the whole affair late in the show's second and last season that, as far as he was concerned, "the CBC was not brought into being to instigate or stimulate social change." It must, he said, "let people make up their own minds" and this required that "it serve public opinion rather than moulding it."[21] *Seven Days* has gone down in CBC lore as a program before its time, a tall poppy cut short by a timid management, a kind of AVRO Arrow of broadcasting.[22] "Hold on to your hats Canada," says the cbc.ca website where you can currently watch a selection of the show's greatest hits, "television is about to change forever."[23] But, at the time, it divided many minds and many loyalties. Eric Koch who would go on to write a history of the program—*Inside Seven Days: The Show That Shook the Nation*—experienced this conflict as an acute personal "dilemma." As someone who had joined the CBC in 1944 and been part of radio's vaunted "golden age," he had been a charter member of the old Talks and Current Affairs department. By the time *Seven Days* went on the air, he was the supervising producer in television current affairs, a job which included oversight of *Seven Days*. He was "greatly impressed," as he later wrote, "by the idealism, courage and vitality of the [*Seven Days*] producers," but he still "in many ways felt closer to the old guard than the young turks"—the old guard being those who, in some measure, shared Alphonse Ouimet's hesitancy in the face of the idea that the CBC should be Canada's conscience and not just its public forum. This left Koch's sympathies more or less perfectly divided, and, since he was "temperamentally unsuited" for the showdown he could see coming, he, in effect, recused himself by resigning as supervisor and joining the staff of the program *Take Thirty* as an ordinary producer. When he wrote about the battle that ensued and that ended with the show's cancellation at the end of its second season, Koch made the argument that I have already cited several times and incorporated into my own argument: that the show's short life signified nothing less than "the end of the first era in [CBC] broadcasting." Up to the time of *Seven Days*, Koch wrote, the CBC "had many of the features of a secular church," its producers a select cadre embodying a sacred trust whose exercise depended on their disinterested identification with a public good that transcended any political or partisan concern. Afterward, this view appeared antiquated and elitist. The church was, so to say, disestablished, and the CBC was left to fend for itself in a broadcasting marketplace in which it could claim no higher calling than anyone else.

I think Eric Koch is right that *Seven Days* marked a watershed between distinct eras in the life of the CBC. It may be true that if Watson and Leiterman hadn't invented *Seven Days*, someone else would have created a comparable uprising under a different name. It's certainly striking that the program that succeeded *Seven Days* in the same time slot—a show called *Sunday* under the direction of Daryl Duke—created no comparable stir, even though it was, if anything, more scandalous. *Seven Days* in this sense, was a sign of the times. But it still changed the CBC decisively—its cancellation, the pyrrhic victory that eased the old guard's retreat, its sacrifice the sanctifying blood that sealed a revolution.

Beginnings are revealing—"In my beginning is my end," says T. S. Eliot in his *Four Quartets*—and I think that Koch's idea that *Seven Days* inaugurated a new era is particularly helpful now that we are at the end of that era.[24] Much can be seen, in embryo, in the styles, devices, and attitudes adopted by *Seven Days*. Where I disagree with Koch a little is on his sense that the CBC lost the character of "a secular church" in its second era. He is right, I think, that the institution itself may have become less hallowed and less able to command loyalty and doctrinal uniformity. But, in other ways, the corporation became more church-like. The old view of the CBC—as a public utility or public square where people, as Ouimet told Parliament, "make up their own minds"—may have been naïve, but it also expressed a real sense of modesty, formality, and self-limitation. The CBC, in its first era, did not present itself as an advocate or representative of its audience. It did not promise justice or vindication, as with Leiterman's imagined "little man," exulting, as the mighty were brought low, "Boy, they really got that guy where it hurts." It did not promote its hosts as heroes, intercessors, or models of sensibility. All this changed in the second era, and, insofar as the word "church" refers to a mediating institution that performs in place of its congregation, the post-*Seven Days* CBC became more church-like in the intensity, scale, and ambition of its mediation. I will argue this in more detail in a moment when I examine some of *Seven Days*' innovations, but first, I want to look at a few of the ways in which the show reflected and enacted its times.

Signs of the Times

New technologies excite millenarian hopes. Television in Canada was only four years old when Patrick Watson joined the CBC, and the new service was just finding its feet. In more recent years, we have seen the extravagant dreams evoked by social media—peaceful revolution, "a more open and connected world," etc.—and the bitter disappointment and disillusionment that has followed, as the shadow of these new techniques has manifested itself in civil war, shortened attention spans, crowd contagion, and monopoly. Television's arc from saviour to destroyer was longer but otherwise comparable. In its first phase, television "stopped a war," as was often said, somewhat extravagantly, of the Vietnam War. It shone its light into dark corners, as the *Seven Days* manifesto promised, righting wrongs and holding politicians to account. In its second phase, well characterized by Neil Postman's 1985 title *Amusing Ourselves to Death*, television was seen as the very bane of dignity, civility, and rationality.

The techno-euphoria of this earlier period is epitomized by the interview with Marshall McLuhan that concluded *Seven Days*' final season. This was a fitting climax insofar as the whole show, as an exploration of "what works on television," had been conceived in the spirit of McLuhan's later maxim that the medium is the message. McLuhan, in a sober mood, was as profound and penetrating a critic of media as the world has seen; but he also had his more oracular and ecstatic phases, and he was in one of them when he sat down to talk to Robert Fulford for *Seven Days* in May 1966.[25] "Television," McLuhan told Fulford, "creates an enormously serious ... realistically minded kind of person—almost oriental in his inward meditativeness." Older media like the movies, he said, generate a "fantasy escape world," while the television viewer remains self-possessed, always aware that "things have effects." I repeat this claim not to embarrass a thinker I hold in high esteem—McLuhan's serious work rises far above the simplistic *x* causes *y* style of analysis into which he was sometimes inveigled, or inveigled himself, during his years of media celebrity—but to show that *Seven Days* was produced in a state that one might call social intoxication. What could television not do in the right hands?

Seven Days' excited techniques mirrored and were mirrored by many other contemporary happenings. There was Beatlemania—*Seven Days* was there at Maple Leaf Gardens with close-ups of adolescent girls melting in ecstasy. There were psychedelic drugs—*Seven Days* visited the commune Timothy Leary and Richard Alpert had established in Millbrook, New York, and found a young adept who had "entered a cavern of dripping jewels" and "become the idea of a stained-glass window." There was a growing anti-war movement to which *Seven Days* contributed with the documentary I mentioned earlier, Beryl Fox's impressive *The Mills of the Gods*; and an anti-psychiatry movement which sent *Seven Days* to the Ontario Mental Hospital at Penetanguishene with a hidden camera to publicize a case of questionable psychiatric confinement. In all these ways, and many more, *Seven Days* reflected, as well as helped to create, the excitement of the times.

Seven Days was also in tune with the times philosophically. This can be seen in the program's challenge to the received doctrine that the CBC must not express an "editorial point of view" or take "position[s] in matters of public controversy," as Vice President Eugene Hallman had put it. Hallman's statement implied a clear and unmistakable distinction between facts and values. But, by the time, *Seven Days* went on the air in 1964, this account of "objectivity" was beginning to wear thin in many places—not just at *Seven Days*. I have already mentioned the influence of Marshall McLuhan who had revived Harold Innis's claim that all media have an inherent bias. Equally significant were the changes in the philosophy of science that were signified by Thomas Kuhn's celebrated book *The Structure of Scientific Revolutions*, published in 1962—a book whose impact far exceeded its readership. Kuhn argued that scientific thought is organized in what he called "paradigms," a word rarely heard outside of specialized scientific milieux until he launched it into everyday use. Then, suddenly, "paradigm shifts" began tripping off all tongues. Paradigms, for Kuhn, are the coordinating frameworks within which facts become facts. A new paradigm will bring to light new facts and make old facts irrelevant. Kuhn was talking about Ptolemy v. Copernicus, and Newton v. Einstein, not Head Office v. *Seven Days*, but the pertinence of his theory was clear enough. There is not one objectivity, but many objectivities, according to one's starting assumptions. The parallel with McLuhan's extension of Innis is also clear. The "message" carried by media is shaped and conditioned by a given medium's underlying physical structure and inherent biases; knowledge is given form by the paradigms which organize it. Both insights might be said to be part of an overarching philosophical revolution which has brought the confident, progressive, world-conquering faith underlying modernity into question in our time. The so-called linguistic turn in philosophy would be another example. Metaphysics was brought back to earth when attention was focused on the slippery and untransparent linguistic tools that had formerly been taken for granted. Objectivity again was put out of reach.

Another contemporary transformation in which *Seven Days* played its part was the change that television was effecting in the practice of journalism. American journalist Lewis Lapham recalled that when he set out on his career as what was then still called a newspaperman, in San Francisco in the 1950s, the sense of something a little bit seedy and disreputable still clung to his new occupation. The power of scandal and exposure "the papers" exercised still gave them a whiff of brimstone in the nostrils of polite society. But this began to change, in Lapham's view, around 1960 ...

> ... you begin to get a whole new kind of person into the press—people who have been to Harvard or Yale or Princeton, people who see the press as a way to achieve fame, fortune and celebrity, which had not been the notion of the older generation in the newspaper profession. Everything is changing. It all fits together because you get the techniques of television and you get the more educated, more ambitious class of people going into the media, and at the same time you have the media becoming a more and more important economic force.[26]

Lapham knew whereof he spoke. Being himself of patrician origins and educated at Hotchkiss, Yale, and Magdalene College, Cambridge, he was both a harbinger and a wry observer of the trend he identified. What he says certainly applies to *Seven Days*. Leiterman had studied at Harvard as a Nieman Fellow in 1954; Watson was fresh out of graduate school at the University of Michigan when he came to the CBC in 1956; host Laurier Lapierre was, among other things, a university professor. There were no graduate degrees at headquarters, just a lot of practical experience. So, *Seven Days*, one can say in summary, embodied a perfect storm—a new medium, a new class of journalists, a new intellectual ethos, a new economy. Television was melding with other social forces to produce the sense that "the whole world is watching"—that resonant chant of the embattled demonstrators at the Democratic Convention in Chicago in 1968.

I am condensing and foreshortening here. Modernity has always had its challengers and countercurrents; journalism was sometimes respectable before 1960; not everyone read McLuhan and Kuhn; objectivity didn't suddenly come unglued when *Seven Days* premiered in the fall of 1964. But my point is that "history," at that moment, seemed strongly on the side of *Seven Days*. Management, in their Ottawa fastness, was obviously out of touch, and the show was popular as no CBC Television program had ever been before or has ever been since. When the news became public that the CBC had decided not to renew the contracts of hosts, Patrick Watson and Laurier Lapierre, Save Seven Days committees were formed around the country, the House of Commons Broadcasting Committee began public hearings, and a fifteen-foot effigy of Alphonse Ouimet was burned by an aroused crowd outside CBC Vancouver. Only now, at the end of the era *Seven Days* initiated, can one begin to see that the objections made by management in Ottawa may have been something more than the reactionary gripes of old radio men whose day had passed. Recovering what was good in the old public service/public utility/"elitist" conception of the CBC will be a task for the final section of this book, when I will try to sketch a theory for a third era of public broadcasting. Here, I want to examine in more detail what was original and prophetic in *Seven Days* approach.

Seven Days as Innovator

Seven Days promised its viewers justice. The program, according to its "manifesto," would "draw attention to public wrongs and encourage remedial action." A good example is the case of Fred Fawcett, an Ontario farmer who was imprisoned at a hospital for the criminally insane in Penetanguishene, Ontario. *Seven Days* became aware of the case early in its first season, when it learned that Fawcett's family believed him sane. A *Seven Days* crew entered the hospital as part of what was said to be a family group, with a portable camera concealed in a picnic basket. An interview with Fawcett was filmed in which he appeared sane, and then the crew more or less ambushed the superintendent of the hospital, Dr. Barry Boyd. He was filmed from his office door and subsequently presented on air in a tight close-up in which his surprise and discomfiture made him look quite ridiculous. Toronto management were distressed to learn of this use of subterfuge and deceit in the gathering of program material but eventually allowed the item to be broadcast after notifying the affected authorities. The case is often remembered as a triumph of investigative journalism—the former editor-in-chief of CBC News, Jennifer McGuire, presented it as such in a 2013 blog post entitled "Half a Century of Investigative Journalism at the CBC,"—and it certainly worked out well for Fred Fawcett. Through *Seven Days*' intervention, the case came to the attention of the Ontario Premier's office and, within the year, Fawcett was released.

But there is another side to the story—the side that worried senior CBC management who feared that their heroic Toronto producers were acting as vigilantes rather than as disinterested ombudsmen. The case itself is tangled. Fawcett had stopped paying his taxes after his township—the Township of Euphrasia in Grey County near Owen Sound—failed to provide an access road to which he thought he was entitled. Township officials were sent to his farm, and he was subsequently charged with threatening them and shooting out a tire on one of their vehicles. He was accused of assault against one of these officials, as well as against a police officer who came to his farm several weeks later. Before he could stand trial, a psychiatric assessment was ordered, and, as a result, he was found to be mentally ill and confined indefinitely at the Ontario Mental Hospital in Penetanguishene. At the time of the *Seven Days* item, four psychiatrists on the hospital's examining board held Fawcett to be insane, while three did not. A legal appeal against his confinement had recently been denied by the Supreme Court of Canada, leaving Fawcett with no further recourse.[27] This may well have been a miscarriage of justice, and one can then fairly credit *Seven Days* for its part in the dismantling of excessive psychiatric power that was underway on many fronts in the 1960s. But questions still arise about the motives of the *Seven Days* producers, as well as about the incipient role of the media as agents of justice.

Seven Days, in seeking justice for Fred Fawcett, put itself forward as the people's tribune, and this was the basis of management's initial objection. Management did not challenge the show's producers on the question of whether or not Fred Fawcett was criminally insane. The issue was whether the CBC ought to put itself above other institutions of Canadian society, acting as their guardian and judge, or pretend to have a role as a court of justice that it could not possibly fulfill. The issue is tricky, as one begins to see when it is framed in this way. Publicity and justice have a complicated and somewhat delicate relationship that can be easily thrown out of balance—too much publicity can be as bad as too little.

Seven Days hardly invented crusading journalism. William Lyon Mackenzie thundered against the Family Compact from the pages of *The Colonial Advocate* in the 1820s. The quotation from Junius that still appears at the head of the *Globe and Mail*'s editorial page—"The subject who is truly loyal to the Chief Magistrate will neither advise nor submit to arbitrary measures"—was put there by George Brown in the 1840s.[28] The Muckrakers exposed issues from which courts and legislatures turned away in the early years of the twentieth century. Wasn't *Seven Days*, in this sense, just CBC Television's journalistic coming-of-age? Yes, in one sense, it was, and to that extent, it became a salutary presence in Canada's political life. But I see another side to the question as well. Can a program that is expressly interested in attracting and holding an audience—Watson's *sine qua non* was, after all, "we got to have them watching"—also promise its viewers justice? Or will it, in the end, only weaken and discredit the institutions of formal justice and then turn away when it loses interest in the less telegenic cases? This question was something that preoccupied CBC management at the time. Their question was, Who will watch the watchers?[29] Television, at the time *Seven Days* went on the air, was still in its infancy, but its potential power to entrain attention and build opinion, as Leiterman and Watson perceived, was immense. Many of the codes by which management operated were intended to keep this power under wraps and the CBC out of the political limelight of which, in their view, it already had dangerously much in the form of frequent denunciations from the Conservative benches in the House of Commons. *Seven Days*, cheeky, charming, in tune with the times, stepped into the spotlight, unafraid to own this power and to trust in its own good will to govern it—remember Leiterman's claim that objectivity mainly consists in the competence and good faith of the person "who uses the tools." But the power of television would not be easily governed, nor would good will always be its governor. And television, as we can now appreciate, was only the beginning.

Radio and television, when they first appeared, were both accounted part of the Fourth Estate—a name for the press that traces back to the eighteenth century, and some say to Edmund Burke. According to this old usage, the body politic is made of three estates—the church, the nobility, and the commons—and the press forms a fourth, standing outside the political constitution of society as observer and critic. When the American Constitution appeared, with its three branches of government, the term was easily adapted to signify the press's independent standing, exterior to this governing apparatus. The Fourth Estate stood apart from the state with its feet in that free creation of public opinion that Jurgen Habermas named "the public sphere."[30] This was the theory, but powerful and monopolistic media soon challenged it. Prescient Oscar Wilde, writing in the first age of newspaper power in the 1890s already saw the problem: "Somebody—was it Burke?—called journalism the fourth estate," wrote Wilde. "That was true at the time no doubt. But at the present moment it is the only estate. It has eaten up the other three … We are dominated by Journalism."[31] Harold Innis, writing in the 1940s, when the mass circulation newspaper remained the dominant medium, carried on this theme. He conveyed a sense of dizzying and disorienting velocity and instability, which might surprise the contemporary reader for whom the daily paper is little more than a reassuring antique. The popular press, Innis wrote, had been responsible for "a ruthless shattering of language." He spoke of "the cruelty of mechanized communication" and claimed that it had led to "a systematic destruction of elements of permanence essential to culture." "Thought," he concluded, "has been paralyzed."[32] One might ask, in the face of these terrifying words, what further harm could *Seven Days*, or anyone else, do? Well, Wilde and Innis were both, in their very different ways, prophets, peculiarly alert to the accelerating rhythm of technology in their times. What they foresaw entered a new stage with the appearance of the very simulacrum of reality on television. Leiterman, Watson, and their colleagues thought that they could ride the tiger, but one can see, at the end of the age they began, that, like the lady in the limerick, their fond hopes have mainly ended up inside the tiger.[33]

Oscar Wilde could already see that the formal and legal constitution of his society no longer corresponded to its actual constitution. In his eyes, the supposedly marginal and compensatory Fourth Estate had already swallowed society whole. But different timelines can be attached to the story of when the anti-establishment media became themselves the establishment. Lewis Lapham, in the passage I quoted earlier, set a watershed around 1960. However the story is told, the end is the same: As the scope of media, and the pretensions of journalists, expand, a power imagined only as a counterweight to existing establishments becomes an establishment itself. But this change doesn't register within the political constitution of society where only administrative, legislative, and judicial power is formally recognized. The construction of knowledge still theoretically takes place outside this constitution in a sphere imagined as a free and unconstrained voicing of the concerns of civil society. Media become a dominant power but remain, it might be said, an unconstitutional one. No one guards the guardians.

Seven Days illustrates. A dramatic case like Fred Fawcett's might draw attention, and even lead to justice, if it were scandalous enough to hold that attention long enough. But maybe the next such case will be less interesting—not quite scandalous enough perhaps, or just lacking sufficient novelty as a story that's already been "done." The point—obvious enough—is that justice and gripping television do not correspond, and, if broadcasters fail to acknowledge the biases introduced by their physical medium, their mode of production, and their professional self-interest, a self-aggrandizing illusion is generated. This illusion puffs up the unaccountable power of media on the one hand, and potentially weakens other institutions that have, or at least might have, justice, not popularity, as their primary aim. All this, I'm claiming, Alphonse Ouimet and his colleagues at head office inchoately and inarticulately knew. But, since they could offer nothing better than a status quo in which the CBC bowed politely to existing social establishments, like the Ontario Mental Hospital, populist arguments about "righting wrongs" and subjecting the mighty to "tough encounters" easily carried the day.

The Host as Hero

A second original feature of *Seven Days* was the role played by its hosts, Laurier Lapierre and Patrick Watson, who emerged from behind the scenes as co-host in the second season. In those days, it was a persistent criticism of the CBC that it tried to discourage the emergence of "stars" among its on-air performers. In his memoir, *The Pierce-Arrow Showroom Is Leaking: An*

Insider's View of the CBC, Alex Barris, a television writer, producer, and host, contemporary with Watson and Leiterman, speaks of the corporation's "antipathy" to creating stars and its lack of "loyalty to artists" and ascribes these tendencies to bureaucratic timidity and fear that the performers might outshine the organization and eclipse its "mandate."[34] This diagnosis has echoed down the years, finding one of its most recent voices in Richard Stursberg, who headed the English radio and television networks of the CBC between 2004 and 2010. Stursberg felt he detected "a profound schizophrenia" at the CBC, which was expressed in the belief that "if [a] show attracted large audiences, it must be vulgar and stupid."[35] But, this criticism points to a feature of the CBC that has, at times, had a more positive aspect. Stars exert gravity, and the bigger they are the more gravity they exert. In the more recent case of radio host Jian Ghomeshi, where the CBC took the opposite tack and amplified Ghomeshi's celebrity, it lived to regret its aggressive and indulgent star-making. If one takes seriously Bernard Trotter's characterization of the CBC as properly a public utility, then this function is best fulfilled when it goes unnoticed, not when it attracts attention to itself. "A leader is best," says Lao Tzu, "when people barely know that he exists … and people say, 'We did this ourselves.' "[36]

With *Seven Days*, a decisive move occurred from the host as connector/convenor/conjoiner to the host as surrogate or representative—someone in whom and through whom the audience could experience its delegated power. There was, for example, a famous incident in which the camera lingered on Laurier Lapierre, as he brushed away a tear during an interview with Doris Truscott, the mother of Steven Truscott, a teenager railroaded into prison for a murder of which he was later exonerated. Management were outraged—CBC President Alphonse Ouimet called Lapierre "unprofessional" and said that he was merely "a good actor," who was unnecessarily trying "to heighten further" what was already an intensely "emotional" atmosphere.[37] Ouimet, then at the end of his tether, was unkind—the pathos of Truscott's unjust imprisonment, particularly when seen through the eyes of his mother, might have inspired tears in anyone—but the point is that Ouimet saw it as the duty of a CBC host to bracket his feelings, while the show's producers felt that Lapierre was acting *for* its audience in expressing a presumptively collective emotion. The host had become more of a priestly figure, acting on behalf of his congregation/audience and not just Bernard Trotter's "neutral," unwilling "to impose [his] own ideas on the country."

This is a highly charged role. At best it may help someone in a bad spot. *Seven Days* got Fred Fawcett out of an institution in which he probably never belonged, and Lapierre's tear contributed mightily to the publicity that led to Steven Truscott's parole three years later, as well as to his eventual exoneration.

But it also invites self-dramatization and a potentially fatal confusion between the interests of journalists and the interests of the public. Consider Ouimet's charge that Lapierre was "a good actor." At the time, it appeared only to denigrate the obvious naturalness of Lapierre's feeling and reinforce the already well-established story that management were unfeeling and out of touch. But, as a CBC Television host, Lapierre was also someone who was trying to convince his compatriots that the tiny selection of photogenic events that could be fitted into a weekly hour of television was, in the words of the *Seven Days* manifesto, "of such natural interest, such vitality and urgency" that it ought to be "mandatory viewing." The CBC, and television in particular, is a place where reality is staged, not where it simply occurs, and the more natural and unaffected the host appears, the more this artificial quality slides out of view.

Eric Koch characterized the CBC, in its first era, as having a church-like character. What I understand him to mean is that there was a sense of vocation—of having been called—and a certain self-denying austerity to the men and women of the CBC's first era. (I'm referring here to a public and political stance, not to private habits.) A common cause engendered unity and obedience. This was partly an effect of the fact that the CBC of the first era was, in a real sense, inventing Canada, by providing a federation of still quite disparate and mostly non-communicating provinces and regions with one of its first simultaneous experiences other than war. It was also an effect of an always potentially hostile political environment in which the CBC's ideological chastity was one of the few trumps it could play against interfering governments. *Seven Days* broke this mould with its political provocations and tabloid style. But there is another way, as I said earlier, in which *Seven Days*' new approach, however flamboyant and unruly, was equally church-like. I see this above all in the show's pretension, embodied in its hosts, to represent and speak *for* the country, and therefore to be its clergy. Lapierre's tear was, in this sense, sacramental, cried for the country of which he was the delegate and model. It was a tear cried to acknowledge and mollify the injustice of Steven Truscott's wrongful conviction. Insofar as a priest is someone authorized to perform a sanctioned rite, this new stance was arguably more priest-like than what had preceded it.

Seven Days' third innovation was its definitive challenge to what Leiterman dared to call the "myth" of objectivity. The word has many meanings, and many more associations, so I will try to say, more precisely, what I intend by "definitive." I certainly don't mean that up until 1964 the CBC was staffed by naïfs who believed they had an angelic power to convey information without distortion and so represent the world just as it is. Objectivity, in that sense, was always a myth. I mean only that, in the first era, the CBC largely saw itself as subservient to an evolving social consensus, which it presented according to commonly understood rules. It was certainly a modernizing, "progressive," and liberalizing influence, but it still minded its manners, addressed its audiences as groups of citizens and strangers freely assembled, and was careful to consult all shades of acceptable opinion. The CBC saw itself as a limited institution within a social order that it served but did not fundamentally challenge. The scale of the change effected by *Seven Days* is illustrated by the following passage from an article in *Canadian Forum* for January 1967. The author, Peter Hughes, speaks of the show's run as a time …

> … when Canadians events briefly took their place in a moral order. Patrick Watson warned us, Laurier Lapierre wept over us, and what was happening here meant something to us because it meant something to them. Both of them seemed convinced that they could govern the country better than the politicians … and each of them carried, if not a marshall's baton, at least a minister's portfolio in his knapsack. The ending of *Seven Days* with its significant overtone of populist outcry and a failed *coup d'état,* was also the end of an attempt to raise the public life of English-speaking Canadians tò a level of interest and moral significance.[38]

Hughes goes on to say that the way in which *Seven Days* heightened the significance of Canadian public life was a particular good because of the powerful pressure the United States exerts on Canadian consciousness, a pressure which tends to undermine our ability to see our public concerns in an ethical light. Particularly striking, for me, is his use of the phrase "moral order." The idea that a television program *could be* such an order sums up the change I've been trying to point out here. Watson and Leiterman clearly aspired to create such an order, as I think my quotations from them earlier make clear, and clearly believed that their good intentions were sufficient to safeguard this order. Alphonse Ouimet and his colleague saw this stance as presumptuous and potentially dangerous to the CBC.

Seven Days asserted itself as its own "moral order" in various ways. It freed television from conventions that were constraining its dramatic power. It heightened audience identification through the mediating and intercessory role of its hosts. It broke the unity of the CBC as a "church" by appealing to the public over the heads of its own management. And it fused the CBC's two mandates—to enlighten and entertain—by promising its viewers justice, vindication, and entertainment in one package. This created a new style which, for want of a better word, I'll call "populist." The word is problematic at a moment when populism is mainly being expressed on the political right.

But, populism, historically, has just as often had a left ward slant, as it did in Canada in the first decades of the twentieth century. And *populism* was the term that was often used at the CBC, at least so long as there was a recognizable "elitism" against which it could define itself. So, I'll retain the word and hope to be understood as referring specifically to issues in broadcasting and not to populist politics more generally.

Populism, as it was formulated by *Seven Days*, was based first of all on Watson's maxim, "We got to have them watching." Watson and Leiterman couldn't save them, if they didn't first tune in. This was a kind of declaration of independence. The old CBC, for better or worse, was socially and politically embedded. It took its standards from other institutions with which it was connected, as well as from prevailing social mores. "Our chief thrust," as Frank Peers said, "was to establish and maintain connections with various segments of Canadian society."[39] A symphony broadcast would reflect the priorities of the orchestra and not just those of the CBC; an academic broadcast would respect academic canons; a farm broadcast would speak directly to farmers without worrying about how entertaining it might or might not be for others, etc. Populism took its standards from its audience alone. To do otherwise was to submit to some elite curriculum whereby the orchestra tells you what to like, the university tells you what you ought to know, or the farm organization tells you what's important to farmers.

Seven Days can plausibly serve as a marker of this change because it incorporated and anticipated so many of the features that would define the era that is now ending. Such a judgment is possible only in retrospect. At the time, *Seven Days* planted its revolutionary seed, most of its producers would probably have seen the status quo as possessing the predominant weight. The CBC didn't suddenly declare that the customer is always right in 1966. Nor did it entirely stop programming for the minority tastes that fell under Reeves Haggan's stricture on "Don Mills talking to Don Mills." It was 2004 before Richard Stursberg took over CBC Television and made Patrick Watson's shibboleth—"we got to have them watching"—an explicit policy by proclaiming the audience to be "the only judges who count."[40] But, even so, I think that it makes sense to see *Seven Days* as the watershed, even if the streams whose course and destiny trace back to it are only now reaching their terminus.

Populism Summarized

Seven Days introduced and embodied a "new paradigm" in CBC broadcasting. And, since my argument in this book depends on this crucial claim—that

Seven Days embodied a vision that was both novel and comprehensive—I will conclude by summarizing the basic features of this new view:

1. The integrity of the organization and its internal standards no longer had priority. *Seven Days* appealed to its audience over the head of its management. If we take Koch's church metaphor seriously, then what Roman Catholic doctrine calls "the sense of the Church," was no longer decisive. The CBC would pursue "the bubble reputation" along with all the other suitors for the public's affection.[41] Popularity became a *sine qua non*—not a possible and pleasant consequence of having done something well, but the very condition of fulfilling "the mandate" at all.
2. Hosts took on a much greater importance—as models, as mediators, as delegates, and as receptacles for audience projections.
3. Distinct publics tended to give way to a less differentiated general public. Groups defined by some interest or competence were replaced by a "consumer" whose "like" or "don't like" constituted a decisive and ineffable judgment.
4. The idea of the CBC as a neutral venue, forum, or facility where citizens could assemble to discuss matters of concern to them declined in importance. The editorial and curatorial role of the CBC increased dramatically. Format determined and defined the "issues" presented by the CBC to a much greater extent.
5. The CBC promised its audience more. The promise of justice was prominent in the *Seven Days* manifesto, but, as time went on, many other things were promised, culminating in today's offer of catharsis, uplift, and constant encouragement. The CBC began to offer professional service. At first this was journalistic service—"drawing attention to ... wrongs and encouraging remedial action," etc.—but it gradually extended more and more into an intimate and personal sphere that would have been considered entirely private in the first era.
6. Reservations about counter-productive effects of media were increasingly set aside. Glorying in the potentials of TV, exploiting fully its dramatic power, and submitting in a very "disciplined" way—Patrick Watson's word—to its requirements and exigencies were all hallmarks of *Seven Days*. In order to fulfill the mandate, and reach the people "where they are," it was necessary to "obey" the medium rather than counteracting it. As the power, penetration, and reach of media increased, as it steadily did, this stance fostered the transition from a society served by media to a society constituted by media.

This is not intended as an exhaustive list but just as a set of starting points for further reflection as this essay goes on. The point that I want to emphasize in conclusion is that the CBC, after *Seven Days*, increasingly undertook to be a model or replacement society rather than the servant of an existing society. "Canada Lives Here," the corporation's catchphrase between 2007 and 2014, expresses this stance. The CBC is a place, an image, a brand, and not merely a facility. Whether it was the more just Canada that *Seven Days* hoped to create, or the cozy, communal Canada on *Morningside*, or the hip Canada portrayed in the person of Jian Ghomeshi, before his fall, on "Q," the CBC increasingly became a surrogate that performed in place of its audience. A given program's format, stance, and style began to bear more heavily on the subjects it treated, while the authority of the subjects themselves diminished. By subjects here I mean persons as well as bodies of knowledge, and by authority I mean that subjects of both kinds have an inherent character and dignity that ought to influence how they are approached. Again, it's a question of degree, and also of distinguishing inherent or deserved authority from established and merely habitual authority. No doubt the CBC, in its first era, was at times excessively deferential to established authorities. The point I want to make here is that the CBC, in its second era, often went to the opposite extreme, allowing restrictive formats and self-dramatizing poses to override the authority of subjects altogether.

To return, once more, to Eric Koch and Alphonse Ouimet's image of the CBC as church-like, one might say that in the second era, the institution became less ecclesiastical, but its broadcasts took on a more liturgical character. In a church context, liturgy refers to those rituals by which the church brings itself to life as a community. Liturgy, in other words, doesn't just reflect some existing state of affairs, it enacts or brings into being what it speaks about.[42] In the context of broadcasting, those things I have been calling format, stance, and style, as well as explicit utterance, constitute liturgy. They symbolically enact, and, in doing so, create, an imagined community. The old CBC, I am arguing, exercised this liturgical function much more cautiously and circumspectly than the new CBC of which *Seven Days* was the prototype. The older model was more responsive to other institutions, more respectful of existing mores, and more likely to efface or minimize its own influence on its presentations. The new CBC was more inclined to see in itself "the moral order" that Peter Hughes found embodied in *Seven Days*. Rather than conforming to the world around it, it would make the world around it conform to its parameters, its technical requirements, and its dramatic possibilities.

This created a paradox. Marshall McLuhan got star billing on *Seven Days*, but his maxim—that the medium *is* the message—was not well understood. Media have *biases*, as McLuhan's forerunner, Harold Innis had said—they bring certain things to light and throw others into shadow. No medium by itself can represent the world in all its "blooming, buzzing confusion."[43] But this is not easily admitted by any approach that makes popularity its prerequisite. Self-awareness is apt to lead to self-limitation, and not to the lionization that *Seven Days* sought by roasting politicians, rescuing mental patients, and weeping over the unjustly imprisoned. Television of such "vitality" and "urgency" as to constitute "mandatory viewing" is unlikely to be television that draws attention to its own biases. It will have an overwhelming interest in doing the exact opposite. It will try to efface the difference between representation and reality. The politician in the hot seat must be seen as sweating because he's been found out, not because he's sweltering under hot lights in an uncomfortable chair in a smoky studio. The medium is made to disappear, even as it is glorified and made to appear as never before.

"In my beginning is my end," says Eliot, but also, at the end of his meditation, "In my end is my beginning." The era that *Seven Days* set moving is now over. The power that Leiterman and Watson thought they could harness has borne down the society they sought to reform and replaced it by a reality so entirely mediated that the very distinction between reality and artifice trembles. We live in what American writer George Trow once called "the context of no context," and the question now is how to situate ourselves on such shifting ground, how to return to earth and reinhabit the world "the media" have swallowed up.[44]

A final note: In stressing the paradigmatic character of *Seven Days*, I know that I have inevitably obscured some of the program's unique and, therefore, quite unrepresentative genius. I went back to some of its shows—in 1986 in the course of preparing a history of the CBC on the occasion of its fiftieth anniversary—and found them just as compelling as Watson and Leiterman intended. But what I have wanted to do here is to try to grasp, through *my* paradigm, the character of the age in which *Seven Days* began. This will lead me, in the end, to a vision of a new era, but first I will turn to what I take to be crucial moments in the unfolding of the age that *Seven Days* began.

3 | Radio Revolution

It is often said that radio suffered a gradual eclipse by television after CBFT in Montreal and CBLT in Toronto went on the air in 1952. Like most such summary tales, this story has many exceptions. Brilliant things were done on CBC Radio during the 1950s—from the virtual repertory company of comic types created by Max Ferguson for his *Rawhide* broadcasts to philosopher George Grant's *Philosophy in the Mass Age* first presented on a series called *University of the Air* in 1959. I have fond memories of *Maggie Muggins*, a daily children's program that enchanted me as a youngster in Port Hope, Ontario, in those years. But there is no doubt that television redefined the CBC, replaced radio in prime time, and set the radio service somewhat adrift and in search of a new vocation. Those "Help Stamp Out TV" buttons that Robert Fulford recalled radio producers wearing reflected a genuine malaise.

When radio was the only game in town, it was conceived primarily as an entertainment medium. The Golden Age of Radio at the CBC is now usually characterized by its serious achievements—by *Citizen's Forum*, or by the dramas produced for Andrew Allan's *Stage* series—but pre-television CBC Radio was still primarily a commercial medium, dominated by revenue-producing popular programs. Remember, for example, producer Harry Boyle's earlier story about what an achievement it was to clear a single evening for non-commercial fare on *CBC Wednesday Night*. The change that occurred, as television became the dominant entertainment medium, was that radio began to be conceived more as an information medium. Information and entertainment are, of course, not mutually exclusive, but I'm speaking of a broad change of emphasis. One of its signs was the adoption of so-called all-news formats by some American stations beginning around 1960.

By the mid-1960s, the CBC had officially recognized this changed environment. This is evident in a report called "Programming in the Public Interest" that the CBC submitted to a government advisory committee on broadcasting policy in 1965.[1] The report argued that a "radio revolution" was underway and that the CBC must join it. Anticipating many of the changes that would be enacted over the next few years, it recognized that prime time in radio was now daytime and recommended that the radio schedule, which was then broken up into many short broadcasts, should instead be comprised of longer "block programs" and more topical local "magazine" formats.[2] The committee rejected many of these ideas, fearing that they would make CBC Radio too similar to private radio, but the CBC began to introduce them anyway.

One of the first signs of change was the appearance, in 1965, of *Cross-Country Checkup*, the national weekly phone-in show that is still on the air today. This broke two taboos: the use of telephone sound, until then disqualified from broadcast on the grounds of its inferior quality, and the use of unedited and unscripted material. Even *Citizen's Forum*, which had sometimes been recorded at public meetings, had always been presented as a prepared and packaged digest of the opinions presented. Both the phone-in, pioneered on *Cross-Country Checkup*, and the phone-out, introduced a few years later by *As It Happens*, were new. In both cases, there was relatively direct unedited contact with sources outside the CBC. The studio no longer defined the sound or the scale of CBC Radio.

The change was not instantaneous, of course, and the harbingers of the new style, which stand out so dramatically in retrospect, were still overshadowed, at first, by the perceived crisis of the old style. According to historian Clay Carter, it was "generally believed" within the CBC, that CBC President George Davidson, when he took over from Alphonse Ouimet in 1968, "considered killing the radio service."[3] Carter's statement is hard to assess—Davidson made no explicit statement to this effect, "considered" implies no definite plan, and, around this same time, CBC publicity materials were already pointing enthusiastically at a "revolution" in progress—but, if Carter is right that this threat seemed plausible to radio people at the time, it certainly indicates that a mood of uneasy apprehension was widespread in the years leading up to 1968.

At the same time, a new sense of vocation was quickly developing. *Cross-Country Checkup* was followed a year later by *The World at Six*, a nightly half-hour news round-up that also continues to this day. *As It Happens* went on the air in 1968, *Radio Free Friday*, with Peter Gzowski as one of its hosts, appeared in 1969. The publicity release for the latter described the program as the corporation's "latest contribution to the radio revolution." The same year, CBC Winnipeg refashioned its 6:00–9:00 a.m. broadcast as Information Radio. The Winnipeg experiment, as the CBC called it, became the template for a reconceptualization of all local programs as sources of information rather than as entertainment. The art of radio documentary flourished with programs now mainly recorded "in the field." A new generation entered CBC Radio.

By the time I began to learn and practice the art of radio documentary in 1971, there were already a number of programs that were built on freelance contributions, including *Ideas*, *Concern*, *Identities*, *Between Ourselves*, *Rule and Revolution*, and *Five Nights*, the nightly, fifteen-minute broadcast after the 10:00 p.m. news where many freelancers, including me, took their baby steps. The range of voices heard on CBC Radio expanded considerably.

From the start, as we have seen, these changes were framed as a revolution. When the term was first introduced around 1965, it referred to a revolution that was changing the place of radio in the media landscape generally, but it soon became, and has remained, the name for a more specific CBC revolution. When Jack Craine was appointed as director of CBC Radio in 1965 it was with an explicit mandate to revive a moribund radio service. *Maclean's* magazine, in a feature article by Sandra Peredo in 1968, posed the issues Craine faced in stark terms. Calling the radio service a "near corpse," and using words like "relic-y," "quaint," and "WASPy," to describe the outdated regime that ruled radio, Peredo proclaimed that "CBC Radio is engaged in a last-ditch push for survival. It's a sort of Operation Boot-strap that follows almost 15 years of zombie-like repetition of what it had always done before TV took over our evenings."[4]

Peredo's picture was overdrawn, but it did indicate some of the dimensions of the challenge Craine faced. His indispensable ally was Margaret Lyons, who would go on to become head of radio herself. She had begun her career at the BBC and then returned to her native Canada in 1960, at first as a radio producer and then as head of current affairs. In a video tribute to Lyons, created after her death in 2019 by long-time CBC Radio host and documentary-maker Karin Wells, Peter Herrndorf, a former vice-president of the CBC, called her "the most important, most influential CBC Radio executive of the last sixty years."[5] She was also, according to one of her successors as head of radio, Harold Redekopp, "a streetfighter." Various other metaphors of combativeness and determination accumulate in Wells' documentary—"a tiny piledriver" (Herrndorf again, referring to her short stature), "a fire-breathing force of nature" and one who "took no prisoners" (Michael Enright), a "Pit Bull" willing to "duke it out" (radio producer Doug Ward), and so on. The nickname by which I first knew her was "the Dragon Lady," a term that no one would dare to use today, with its double stereotype of the Asian and the female, but which was then deployed affectionately as part of her considerable mystique.

Margaret Lyons hired many of the producers who shaped the new CBC Radio. Among them were Mark Starowicz, Richard Bronstein, Colin Macleod, and Michael McEwan. All were independent-minded; some had been fired from newspaper jobs for showing it. Lyons was determined, in Starowicz's recollection, "to introduce journalistic values." Robert Harris, another veteran of the time, says that she "wrenched" CBC Radio out of the "artsy/cultural frame" in which it had previously operated "and put it into a current affairs political frame." She "invaded the arts." The tone of CBC Radio grew more brash. Barbara Frum, in her memoir of her years hosting *As It Happens*, writes that she understood her mandate as being "to break all the old safe, CBC rules. The orders were: provide an iconoclastic, zippy, nightly information package. Speak to everybody, not just to the few who genuflect to the sound of a mid-Atlantic accent."[6]

Revolutions need enemies, and straw men make ideal enemies. Frum's conceit that CBC Radio had previously spoken to "the few who genuflect to a mid-Atlantic accent" echoes Reeves Haggan's earlier caricature of CBC Television, on the brink of the *Seven Days* upheaval, as "Don Mills talking to Don Mills." Harold Redekkop, in the tribute to Margaret Lyons I mentioned earlier, remembered that her resentment focused on those who had "come up the easy way"—well-heeled, well-mannered men in blazers who had been to Trinity College was how Redekkop epitomized the type for whom Lyons "didn't have much time." I mention this, not to deny that Margaret Lyons came up the hard way, or to suggest that she didn't now and then cross swords with complacent men in blazers, but because it's so easy and so convenient to turn the past into a cartoon. (Readers may recall that both Patrick Watson and Douglas Leiterman of *This Hour Has Seven Days* had blue-ribbon educations.) The populist revolution that Lyons and her many allies were making needed a certain elitism against which to define itself, and this sometimes tempted them to parody the *ancien régime*.

Populism versus elitism was the binary that framed a lot of debate at CBC Radio in the 1970s, and, for that reason, I will continue to use these terms. But, at the same time, I would like to put them into question. As I said earlier, the populism of CBC Radio producers was of a peculiar kind—it was not the populism that is provoked in an outraged or oppressed people, but the populism of a privileged professional class that wanted, as Barbara Frum says, "to speak to everybody," and, I would add, *for* everybody. Margaret Lyons gave this populism a heartfelt definition in a letter she sent to scholar Clay Carter when he was preparing a history of this period. She had wanted CBC Radio programs, she said, to give voice to "the deeply felt yearnings of their audience." The CBC, she continued, should "lead and mould Canadian thinking" while at the same time bringing people "flashes of instant recognition of *their* insights."[7] This was the same hope *Seven Days* producer Doug Leiterman had expressed when he imagined his ideal cab driver or checkout clerk saying, with satisfaction, "That's what I think!"

The CBC, in one way or another, had always found its vocation in holding up a mirror to Canada and, in that sense, representing it. The first chairman of the CBC, Leonard Brockington, gave full throat to this ambition in the address with which he inaugurated the new service in 1936. He invoked "men in lonely places ... the farmer, the fisherman, the fur trader, rich and poor, great and small, Canadian men and Canadian women" and hoped that all would "share with their fellow citizens the perennial marvel ... of their struggles and achievements."[8] But the radio revolution gave a new twist to this representation. The CBC now presented itself as the people's delegate and not just their mirror. You can hear it in Margaret Lyon's ambition to "lead and mould" as well as to reflect. A report prepared by radio producers Peter Meggs and Doug Ward in 1970 had taken the view—this is historian Clay Carter's paraphrase—that CBC Radio should "help shape Canadian society" and "help prepare the citizenry for social change."[9] CBC Radio would follow in the steps of *Seven Days* with its "probing [of] dishonesty and hypocrisy," its "tough encounters," and its promise of "remedial action" against "public wrongs."

As It Happens

One of the most-celebrated venues of this new, leading-edge CBC Radio was the nightly ninety-minute newsmagazine *As It Happens*. Its philosophy was articulated by its producer Mark Starowicz in a column called "Mediacrities" that he wrote for *The Last Post*, an "alternative" publication of which he had been one of the founders.[10] Writing in 1973, under a pseudonym that protected the CBC, and his situation there, Starowicz claimed that "a major shift has hit Canadian journalism." There had been, he said, "a change in the mood of the society," which had produced "a fad ... for investigative journalism." He offered a number of American examples: Ralph Nader's "success" in bringing to light dangerous practices and complacent regulation in various industries; Seymour Hersh's exposés of Vietnam atrocities in the *New York Times*; and the reporting of Carl Bernstein and Bob Woodward on the Watergate burglary.[11] A telling example of this new mood, Starowicz said, was the then wildly popular movie *The Sting*, a story about two con artists in which the con is so multilayered that you're never quite sure who is conning who. "The average Canadian," he went on to say, "... knows he is being screwed. And what sells is something that tells him he is being screwed and why ... People are feeling robbed. They want the thieves nailed." This new situation, Starowicz wrote, calls for "cynicism," a stance "to be applauded these days." He then commended *As It Happens*' "success ... in nearly doubling its audience, while growing progressively more skeptical," but claimed that the CBC generally was failing in this regard. "While the Syndicate is running the show," he said, "the CBC, figuratively, is exposing litterbugs." (This was an attack on consumer shows like *Marketplace* with its exposés of faulty products, and on what the CBC then called "grievance programming" generally.)

This short essay of Mark Starowicz's, with its unpacking of the populist turn at CBC Radio, took on considerable importance for me in the later 1970s, when I first began to study and write about the CBC. At the end of 1976, I had been fired as one of the co-hosts of the CBC's 6:00–9:00 a.m. program for the British Columbia region, *Good Morning Radio*—an incident I'll describe more fully later in the book. Regional and network management both felt that my conduct as a program host, and, before that as *Good Morning Radio*'s executive producer, had been "too political" and too outspoken to allow me to continue as a member of the CBC's staff, though I was encouraged to continue as a freelance contributor. The incident left me with a keen interest in finding out where the boundaries of acceptable discourse at CBC Radio actually lay. *As It Happens*, in the later 1970s, when it was co-hosted by Barbara Frum and Alan Maitland, was one of my object lessons in acceptable *iconoclasm*. What Starowicz calls cynicism, and then skepticism, is clearly one of the keys. The philosophers who were called Cynics in ancient Greece linked detachment to the practice of virtue, but, in modern times, the term has come to mean detachment without virtue—a standing back merely, and a disposition to disbelieve. Cynicism, insofar as it indicts the venality of all motives, is unthreatening. No positive program is proposed, no hope held out—the listener's pleasure and consolation is to share in the knowingness of the wised-up as "the thieves are nailed."

As It Happens perceived itself as edgy. Pamela Wallin was a producer on the show in those years, before going on to a well-publicized career as a television host, Canadian consul general in New York, and Conservative senator. She recalled, in the documentary tribute to Margaret Lyons that I have been quoting, that the staff of the show were "a bizarre group" who were the "antithesis" of the received image of "a public broadcaster." "They did," she said, "crazy, radical wild things." And yet, however louche the staff may have felt themselves to be, the show mainly stayed within the boundary I had transgressed. (Max Allen's influential documentary "Dying of Lead," which attracted a lawsuit and a court injunction constitutes a significant exception.) My explanation is that the show offered reassurance at the same time as it offered a frisson of fearless radicalism. Important, in this respect, were the personas of the two hosts.

Alan Maitland was a staff announcer and a master of the sonorous style that had given the CBC its sound, and its reputation for stuffiness and formality, in the time when program hosts had to be selected from the "announce pool" rather than hired as freelancers. At *As It Happens*, he deployed this style in a somewhat arch or ironic vein, undercutting it and exploiting it at the same time in a witty pastiche. He performed the cheeky scripts written for him by the show's producers with panache, while allowing his listeners to stand at an enjoyable remove from what he was saying. It was a form of "camp" perhaps—the announcer impersonator—something that is what it isn't and isn't what it is at once. The effect was satire with its sting artfully pulled—an intimation that this was not quite serious.

Beside him sat Barbara Frum—by now a sainted figure at the CBC with her own shrine in the atrium of the Broadcasting Centre in Toronto, but then a relatively new kind of presence on CBC Radio. Frum was an adroit interviewer—shrewd, withheld, and provocatively non-committal. She had a good sense of timing; she was unafraid to play devil's advocate; and she was single-minded in shaping interviews to her own and the program's advantage. But at the same time, there was a hint of the ingenue, of someone somehow innocent of all the worldly, political business that her interviews treated. It came out in her demure, almost girlish giggle, and in the by-play with Maitland that often ended the broadcast. This created the effect that Tom Wolfe called "radical chic" in his book of that name—the effect of encountering members of the Black Panthers at a chi-chi party at Leonard and Felicia Bernstein's penthouse apartment on Park Avenue, the circumstance Wolfe's book describes. The hosts of *As It Happens* were skillfully playing the parts that had been written for them by their "crazy, wild, radical" producers, but they were also signalling that this was a game, an entertainment, and a pose.

I enjoyed listening to *As It Happens* in those years. The show was witty, concise, informative, and smartly paced. But it was, finally, a triumph of format, a term I borrow from Paul Goodman (1911–1972) an American novelist, poet, and essayist, who wrote extensively on media.[12] "Format," Goodman says, "is speech colonized [and] broken-spirited." The point needs qualification, as all things have form, and all messages require media, but Goodman means something more by format than form. Format is the triumph of occasion over utterance. It is speech tethered so tightly to its instrumental and pre-scripted purpose that the possibility of surprise or spontaneity has disappeared. Where format rules, even the most sparkling utterance will be harnessed to this underlying instrumentality. To go back to Starowicz's article: If the producers of the program have determined that their desired listener "knows that he is being screwed," and that they, therefore, should oblige this imaginary auditor with "something that tells him that he is being screwed," then the program will be designed to meet this objective, and all occasions bent to this purpose. This instrumental attitude would intensify at CBC Radio in subsequent years, as people began to design broadcasts for "target demographics," but its seeds were already present at *As It Happens*. Populism at CBC Radio was not about letting the people speak, but about speaking *for* the people. The difference is profound. In the first case, broadcasting is conceived as an adjunct or supplement to a society whose members can speak (and think) for themselves; in the second case, the broadcaster becomes a replacement or substitute for that society which, accordingly, begins to find its proper existence overshadowed by the representation of that existence.

To return for a moment to Goodman's conception of format, his friend and editor Taylor Stoehr summarizes Goodman's thought as follows: Media intensity, beyond a certain degree, brings about "a new social order" in which "the popular imagination," so lively in "the folk traditions of the past," comes to be characterized by "an extraordinary passivity." In this new social order, media engage in "the continual creation of superficial tensions and anxieties" that are then "drained off and relieved without being purged." This might also serve as a definition of addiction—the fix only postpones the need for the next fix—there is no "purgation." Media are a "perpetual motion machine," Stoehr's paraphrase of Goodman concludes, "designed to run a consumer economy."[13]

Ideally, the CBC, as a public broadcaster, should stand outside the commercial imperative requiring it to continually feed this machine, but, in practice, such independence has always been out of reach. The CBC has been forced to be commercial in both the literal and the more extended sense of the word. Literally, it has been commercial because its primary popular vehicle, first radio and then television, has always been dependent on commercial revenue. But it has also been commercial because it has at all times been engulfed by a commercial environment, which lends nearly irrefutable force to the idea that public service is meaningless without popularity. Richard Stursberg, the head of English Television between 2004 and 2010, put this second point bluntly in his memoir of his CBC years: "The *only* relevant criterion [of a television program]," he wrote, "was whether Canadians watched."[14]

As I showed earlier in the case of *Seven Days*, *As It Happens* was in thrall to contradictory imperatives—popularity and public service—which it wanted to claim were harmonious. *Seven Days* had asserted that it could serve justice and become "mandatory viewing" in the same gesture. *As It Happens* made a similar boast: The "thieves would be nailed" and the show would double its popularity at the same time: skepticism would lead to good ratings. But perhaps what really counted was the "zippy ... package," as Barbara Frum put it. Perhaps the stance of cynicism was assumed, as Starowicz admitted, because "what sells," in his words, is what confirms the current "mood of the society." A certain ambiguity is inherent in the populist stance. And one can also ask whether "the society," presuming there is such a thing, has only one mood, or many. This question will recur. Here I just want to point out that this idea of a singular social mood—sometimes also referred to under the name "the culture"—has become more and more salient at the CBC in recent years. One index of this salience is constant recourse to the word "we" with its implication that the broadcaster typifies a homogeneous sensibility and speaks for a unified body of opinion. *As It Happens* gave this tendency one of its first and most ingenious expressions.

A word more about Paul Goodman's conception of format and its distinction from form. Form is what all things must possess in order to stand up and hold together. Format is this property in such excess that it stifles what it purports to organize, forcing what should stand up to sit down. But how is the right amount of form to be determined? My view is that the right amount of form can only be thought of as existing on a continuum and in a context. Goodman thought that "a new social order" had come into existence and that it was characterized by a gross imbalance between the self-conscious and prefabricated speech of the media and the spontaneity that had once been expressed in the unmanipulated and unmanaged voice of vernacular culture, or folk tradition. People had lost their voice, Goodman thought, because of the predominance in their talk of words and ideas concocted in an institutional sphere and then served to people as if they were home cooking. A necessary balance between the institutional sphere and the vernacular sphere, between prescription and spontaneity, between instrumentality and gift, had been lost.

It is in this context that I think *As It Happens* innovations should be assessed. No argument for an enclosed or romanticized vernacular community is implied. A great good was done, as Orville Shugg said earlier, when the CBC Farm Broadcast reached into the life of rural Nova Scotia with accurate livestock prices and put the itinerant drovers who had exploited the farmers' ignorance out of business. In that case, the balance had to tip more toward the institutional sphere. But there is a point when media as representatives of this sphere begin to exercise an unwholesome hegemony—replacing and supplanting the society they ostensibly, and often ostentatiously, serve. It is on this continuum that I want to locate *As It Happens*—as a moment in a continuing story, much bigger, finally, than the CBC.

Sunday Morning

The perceived success of *As It Happens*, and of Mark Starowicz as its presiding genius, led in 1976 to the creation of *Sunday Morning*, the radio revolution's most elaborate creation, and, by many accounts, its capstone. "When *Sunday Morning* went on the air," says historian Clay Carter, "a keystone was tapped into place. A circle had been closed, and the radio revolution was over." *Sunday Morning* was a triumph of format. The fifteen-page document that was circulated, in advance of its launch, to freelancers wishing to contribute to the show, prescribed in detail "the style and method" they were to employ.[15] The show strove for a "television on radio" you-are-there effect. "We are filming events in sound," the style guide advised. If you can't "GO THERE," "MAKE IT SOUND LIKE YOU ARE THERE." The proper tone—"highly actualized"—and sentence structure—"no subordinate clauses"—were specified. A lot of aphoristic rules were proposed: "A dislocated voice in a quiet room is boring radio;" a documentary must be a "properly framed, composed and planned event," not something discovered after the fact in "miles" of indiscriminately recorded tape; avoid data because "radio is most efficient in conveying 'temperature,' scope, anger, bitterness, texture." An exacting discipline was to be exercised in commanding and holding the listener's fitful attention—"statistics are not absorbable while you're washing the dishes in Mississauga. People pee in the middle of your beautifully constructed passages. Phones ring." Under these circumstances—a disposable product bidding for the attention of a fickle listener—"talk is our enemy," the guide says. Everything must be reduced to some telling moment, some epitome or quintessence of a situation's meaning.

Sunday Morning typified the "journalistic values" that Mark Starowicz credited Margaret Lyons for introducing to CBC Radio. The people were to be informed, and the discipline necessary to capture and hold their sometimes capricious attention was to be exerted. Such an exercise, often enough, has an element of elite condescension. From Patrick Watson, with his PhD in English literature to Richard Stursberg, with his noted art collection, those who have set out to scourge elitism at the CBC have often been themselves elite. This might lead to a facile charge of reverse snobbery—a charge sometimes laid against Stursberg when *he* was trouncing elitism during his time at the helm at CBC Television—but I'm more inclined to ascribe it to a certain asceticism in pursuit of a mission. "If not audience, what?" wrote Stursberg tersely in his book on his CBC years.[16] The people have to be reached and accommodated where they are, not where they ought to be; the attention span they actually have, not the one they ought to have, must be taken into account. *Sunday Morning* was a hit; and, later on, Stursberg would cancel programs that were reaching between 200,000 and 300,000 viewers and install others that often reached over a million.

But there were costs as well. Im*media*cy via a *medium* implies an attempt to disguise the mediation that is, in fact, occurring in order to preserve the illusion. The *Sunday Morning* style guide urged careful planning and premeditation—the program ought to be "largely edited" before it is ever recorded—but it must be presented as if it were happening before your ears-turned-eyes, according to the *Sunday Morning* technique of "filming ... in sound." Many other things must be taken for granted. The more you wish to connect *immediately* with your listener, the less will you be able to bring your own techniques and standpoint into question or even into view. The THERE where YOU are in YOU ARE THERE must be presented as the only place to be—the place where all eyes are currently turned. Walter Cronkite's famous nightly sign-off on the CBS evening news—"And that's the way it is"—encapsulates the perspectival trick on which all news is based—that somebody authoritatively knows what it is. With "journalistic values" goes the confident assertion that the journalist knows at which joints the world should be carved up, and the assurance that this knowledge belongs to the journalist as a professional qualification. The criteria by which the news is selected, the standard against which importance is measured, the sense of when a story has suddenly ceased to be a story—all this must be ascribed to instinct and left off the table—exempt from serious critical scrutiny. Story selection will of course be discussed, but always within parameters that are not discussed. The sense of excitement that is to be generated in the listener must take place on terms that can be taken for granted—the terms broadly speaking of the status quo. For YOU ARE THERE to work there must be the assumption that YOU already understand why the subject is significant and what this significance is. The desired and carefully executed impression of breathless immediacy would dissolve if any critical distance were permitted between the program, the subject, and the audience.

In the final section of this book, I will try to flesh out in more detail what might be called a critical practice of broadcasting. In such a practice, both the biases of the medium, and the techniques by which broadcasts are produced, would be allowed into view and become questionable. The CBC would then no longer speak *for* its audiences, in a mysterious gesture of identification, but would become a vehicle through which the disparate and often divided communities that compose the country could potentially speak for themselves. In the scheme I've borrowed from Eric Koch, in which *This Hour Has Seven Days* inaugurated a second era at the CBC, this would be a third era in which certain features of the first era would be retrieved on a new ground. What I have tried to do here is to establish some of the features of the populism that began with *Seven Days* and was elaborated in the radio revolution. "The owl of Minerva spreads its wings only with the falling of the dusk," Hegel wrote, meaning, among other things, that where something is going cannot be fully determined until it reaches its end.[17] The end of populism, CBC-style, is now at hand, and this allows discernment of certain tendencies within this populism that may not have been so salient at the time that the radio revolution was triumphantly sweeping away the old regime and sending "Don Mills" and "the few who genuflect to the sound of a mid-Atlantic accent" to the guillotine.

Other Aspects of the Radio Revolution

As It Happens and *Sunday Morning* expressed only one facet of the radio revolution, and, crucial as it was, I don't want to overemphasize it. Excessive formatting reduces diversity by imposing a house style, but the late 1960s and early 1970s also saw a big increase in the diversity of voices on CBC Radio. As I mentioned earlier, many new documentary programs were created, among them *Identities*, *Concern*, *Between Ourselves*, *Rule and Revolution*, and *Five Nights*, and already existing programs, like *Ideas*, expanded their scope. This created a lot of opportunities for freelancers, and, even allowing for the conventions of each program, a considerable variety of styles, sounds, and subjects. I got my own start contributing to almost all of the programs I have named, and remember myself, and others like me, having had a remarkably free hand in the creation of our shows. The radio revolution, in this sense, like other revolutions before it, was a soup of many flavours, a *minestrone* able to incorporate whatever ingredients were at hand. A tendency had been created, but more inflexible recipes, and prescribed styles, took over only gradually.

Another telling expression of the genius of the radio revolution was *This Country in the Morning*, which ran from 1971 to 1975, with Peter Gzowski, for most of that time, in the host's chair. *This Country* eventually gave birth to *Morningside*, to which Gzowski returned in 1982, though not before the program went through a number of less acclaimed permutations.[18] Both *This Country* and *Morningside* gave fresh and vibrant expression to the new spirit of Canadian nationalism that was then taking shape in the arts, in politics and in many other departments of Canada's public life. Gzowski had an infectious enthusiasm for this new spirit, as well as being himself one of its embodiments, and this imparted a sense of adventure and discovery to these programs. But what I particularly want to draw attention to here is the intimacy that Gzowski developed with his listeners. Radio is intimate by nature.

Seeing allows distance—we know where the objects of our vision stand in relation to us. Sound, unseen, encloses and enfolds us. Broadcasters have always known this. CBC Chairman Leonard Brockington invoked "men in lonely places" in his inaugural address over the new network in 1936. Women, in isolated situations, wrote to *Trans-Canada Matinee* in the 1950s to say that the program, and the company it provided, was a lifeline.[19] The disembodied and dislocated voice lends itself to imagination, and listeners form close and occasionally unnerving bonds with the voices coming out of their radios. Gzowski's case, even so, was singular.

The executive producer of *This Country in the Morning* was Alex Frame, who would go on to be the head of radio between 1998 and 2002. Frame had been hired a few years earlier by veteran CBC producer Harry Boyle. When *This Country* was being planned, Frame recalled, the older man gave him some salty and somewhat surprising advice: "Women are horny in the morning," Boyle told his young apprentice, "and they're looking for a 'relationship' with their radio."[20] Whether Gzowski or Frame ever consciously set out to take advantage of this supposed disposition, or even believed that it existed, is more than I know, but I often heard from *Morningside* producers about the remarkable number of the program's female correspondents who seemed to be in love with Peter Gzowski. Whatever one makes of this, it seems undoubtedly true that Gzowski created a remarkable bond with his listeners. He was a complex, perhaps somewhat driven man, and on the only occasion I worked with him he treated me with a coldness amounting to discourtesy. But on the radio, his warm, inviting presence and his boyish, slightly stammering manner seemed to take his listeners deep into his confidence. His enthusiasm for many facets of the country's life, from politics to sport, literature to popular culture, generated a vision of what Martin Luther King called "the beloved community." People felt included in this charmed circle through Gzowski's easy kinship and rapport with his contributors and guests—be it W. O. Mitchell or Wayne Gretzky, Jan Arden or Margaret Visser. Canada as an "imagined community" became real.[21]

This Country in the Morning was the beginning of this adventure, and I can remember the sense of homecoming that I sometimes felt in listening to it. Later, I would reflect more critically on the relationship between Gzowski and his audience. There is a continuum in broadcasting between formality and intimacy, between standing back and reaching out, and one way of defining the second era in Canadian public broadcasting would be as a full traversal of this continuum. At its beginning, to use the stereotypes of the *ancien régime* to which I have recurred, stands Reeves Haggan's "Don Mills talking to Don Mills" and Barbara Frum's "few who genuflect to the sound of a mid-Atlantic accent." At its end is the current CBC addressing its listeners as like-minded familiars separated by no social, intellectual, or political difference whatever—"Canada lives here." The remarkable communion Peter Gzowski achieved with his listeners was a station on this way. An age shows features, in the light of its ending, which were not so salient at the time they were first expressed. Peter Gzowski's mode of address to his audience did not define an epoch for him and his producers as it now does for me. They were focused on the Canada they wanted to evoke and serve, and whatever brought them closer to its inhabitants was taken as a good.

What appears to me now is that a dissolution or collapse of the distance between broadcaster and listener was taking place, and one which has finally disabled the CBC as an instrument of critical thought. It is quite unfair, in one sense, to in any way link the sunny, expansive charm of *This Country in the Morning* with today's predicaments. But prospect and retrospect are the same only in heaven, and retrospect will always discover tendencies that were evident to few at the time. Broadcast populism wanted to please; it wanted to change the world for the better; and it wanted to carry on as it had begun. When the managers of CBC Radio began to feel that the revolution's momentum was flagging around 1990, they invented "Creative Renewal" in order to reinvigorate it. Bronwyn Drainie, who had been the co-host of *Sunday Morning* when it first went to air, called for a "second radio revolution" in the pages of *The Globe and Mail*.[22] Producers were encouraged "to break new ground"—the revolution must continue. A point of balance on the continuum I have described was never sought—the spirit was always of pushing on to the end. Alex Frame, when he became the head of radio at the end of 1998, actively fomented "the second radio revolution" that Bronwyn Drainie had called for. "We shook up radio even though it had 95 percent approval ratings," he said later, "because when you have 95 percent approval ratings, you are the status quo."[23] It is my contention that populism has now reached the terminus to which it was being so earnestly pushed.

Resistance

Successful revolutions write their own history, but, even during the time when the radio revolution was in the ascendant, there was considerable resistance at the CBC. In 1972, William Young, the first producer of *Ideas* along with Phyllis Webb, intervened at a CRTC hearing on CBC Radio to warn that the pursuit of popularity might cause the CBC to lose sight of its mandate. Young, who by then had left the CBC, argued that the CBC, by courting popular "success," risked becoming a prisoner of that very success.[24] A few years later, a group of disgruntled arts producers mounted what came to be called "the producers' revolt." It began when a committee of arts producers that was considering "the future of Radio Arts" bypassed their department head, Robert Weaver, and went directly to the president of the CBC with complaints that arts programming was being slighted and was consequently declining in both quantity and quality. They also expressed concerns about the way in which popular culture was beginning to predominate over and displace more traditional arts on CBC Radio. A member of this committee, producer John Reeves, subsequently claimed in an internal memo to his colleagues that "the whole future of good programing is threatened." The essential issue is summed up in the remark I quoted earlier from producer Robert Harris: Margaret Lyons had "wrenched" the CBC out of its "artsy-cultural framework" and put it into a "current affairs political framework." As a result, as the majority of the arts producers saw it, the current affairs department had taken centrestage, depriving the arts department of resources and attention and pushing it to the margins.

The arts producers' action gained favourable attention from Toronto's newspapers. "The CBC is losing sight of its objectives," said a *Globe and Mail* editorial. "Let those who believe that news must be made more fun, that Canadian drama is too heavy for the masses, or that anything serious or inspiring has no place on the airwaves, withdraw to private broadcasting where they belong."[25] Blaik Kirby, writing in the same paper, declared "the new programs" on CBC Radio to be "empty, ephemeral, and without a trace of skill or content."[26] Dennis Braithwaite in *The Toronto Star* said much the same, though he at least made the supposed inanity of CBC Radio part of a more general trend: "We're all getting dumber every day," he asserted. "The media rationalize their leadership of the rush to mediocrity as a realistic or even democratically high-minded policy of giving people what they want, instead of what somebody says they should have."[27]

These testimonies have to be taken with a grain of salt—newspapers, in those days, generally favoured a "serious [and] inspiring" CBC which didn't encroach on their turf as popular media—and yet I think they do indicate that there was considerable disaffection both within and outside the corporation. In the end, however, the producers' revolt was only an inconsequential skirmish—a rearguard action to cover a defeated army's retreat. The style of programming the arts producers had wanted to reinvigorate persisted, but only as a remnant. Robert Weaver's *Anthology* continued to present new Canadian writing for another ten years. John Reeves never bent an inch in his devotion to the "good programming" he had declared threatened in 1977. When I knew him in the 1980s, Reeves was producing a series called *Celebration*, which presented chaste, formal, and serious-minded programs, often composed around Christian texts and musical settings. He was undisturbed, so far as I know, in his out-of-the-way niche on the FM network, but after his retirement as a producer in 1987, nothing remotely like what he did was heard again on CBC Radio.[28] Populism had carried the day.

I use the term "populism," as I've explained, because populism versus elitism was the way in which what was at issue at CBC Radio in the later 1970s was commonly framed. Both camps pursued "the mandate" as the purposes enjoined by successive broadcasting acts have always been known at the CBC, but by different paths. The populists presented their courtship of public affection as a discipline that must be assumed, a sacrifice that must be made, in order to fulfill the CBC's mission—"We got to have them watching," "If not audience, what?" and so on. The elitists argued, as William Young told the CRTC, that popularity would sooner or later become a prison in which the CBC would become captive. For them what constituted "good programming," in John Reeves's phrase, rested on moral and aesthetic judgments that could not be reduced to what the audience, at a given moment, might want. These competing views were, in one sense, just the latest expressions of the divided mind that had been inherent in the CBC's existence from its infancy. The CBC was born, and has lived, in a cultural and political milieu in which "the market" provides the primary index of value. In such a society there is no higher standard of who is fit to rule than the decision of the electors, and no superior measure of what something is worth than the price people will pay for it. But the CBC is also enjoined by law to "enlighten" Canadians—this is the word used by the 1991 Broadcasting Act—and enlightenment is presumably something more than whatever is currently in highest demand. The tension is constitutive and, consequently, structural. The CBC should give people what's good for them—enlightenment—and what they want—entertainment. The equilibrium was bound to be unstable, even while it lasted.

On the other hand, these two approaches are only ideal types to which no one ever wholly conforms. Taken on their own terms the elitists are as eager to please as the populists, and the populists as eager to serve as the elitists. Snobs can be found amongst the populists, unpretentious, and down-to-earth sensibilities among the elitists. In the years before his retirement in 1987, Bob Weaver, the producer of *Anthology*, one of the emblems of old-school arts programming, was my favourite conversational partner on the subject of baseball and the fortunes of Toronto's team, the Blue Jays—a subject on which he was passionate and learned.

A False Dichotomy?

The difficulty I would like to pose lies with the dichotomy itself, and the assumptions that underpin it. First, there is a problem of definition. Suppose you are an aficionado of folk music—you would currently find almost nothing that suits your taste on CBC Radio. Evidently, you are part of a subculture that the CBC lacks the inclination, or the resources, to serve, and yet folk music is a quintessentially popular art, and once the very voice of the people. The same stricture applies in the opposite direction. CBC Radio's current devotion to popular culture often borders on nerdiness, when viewed from the perspective of someone who doesn't mistake "the culture" for the current market in songs, movies, and celebrities. I still recall my amazement, when Jian Ghomeshi's '*Q*' took over the morning period once ruled by Peter Gzowski's relatively catholic nationalism, at discovering how fantastically discriminating and nuanced a knowledge of pop music the listener to '*Q*' was expected to have. At elitist *Ideas* it was a matter of pride and principled professional courtesy not to make such assumptions about our listeners with regard to the subjects we treated.

The populist/elitist dichotomy is also unsatisfactory in taking the character of *the public* for granted. One speaks of the public's tastes, needs, faculties, and even "yearnings," as if these were all definite and ascertainable quantities. Disagreement is limited to the question of how this known quantity is best served. But this leaves out a crucial dimension of the public—a dimension that cannot be captured by the statistical methods of audience research. The public, as well as something known, is always also something imagined, something not yet known—something yet to be created and, therefore, still out of reach. Broadcasting, in this constructive or imaginative vein, *makes* publics. People discover affinities they hadn't known they had, see things from new perspectives, and think thoughts that had not previously occurred to them. They find themselves summoned and assembled in ways not predictable in advance. Publics, in this sense, are evoked, rather than served; called rather than catered to. They cannot form a "target audience" or "target demographic" because they have not yet heard the call that will make them a public.

Populist rationales for public broadcasting have generally rested on service to the taken-for-granted public—Doug Leiterman's "cab drivers and checkout clerks," Mark Starowicz's "average Canadian [who] knows he is being screwed," the easily bored "little old lady in Calgary," who was invoked by my very first producer at CBC Radio, when I didn't want to eliminate a program element that she found tedious (and I thought crucial). This authorizing public—the general public—has now ceased to exist as a unitary and homogeneous entity, along with many other related, and equally formative, modern myths.

Terms that once seemed to point at compact, consistent, and integral ideas—terms like society, science, development, or progress—now indicate tangled knots of intense in-fighting. The CBC was born into the era when these ideas commanded wide assent, and society could still be conceived as a coordinated "social system."[29] The populist/elitist dichotomy was plotted on this implied background grid. This order has now ended, but news of its demise has not yet reached many of the combatants. The case is rather as it was with the fierce Battle of New Orleans, fought between British and American forces in January of 1814: The battle took place eighteen days after the Treaty of Ghent had formally ended the war, because word of the peace had not yet arrived at the mouth of the Mississippi. What we need now—and might have, if news of the war's end reaches the CBC in time—is a forum for discussion whose premise is peace, whose method is free inquiry, whose scope is catholic, and whose aim is the reconstruction of a viable way of life. The CBC was made for such a purpose. The owl of Minerva flies at twilight.

4 | Creative Renewal

"Once more into the fray" was the headline in *The Globe and Mail* under which former Sunday Morning host Bronwyn Drainie issued her call for "a second radio revolution" in early 1990.[1] At the time she wrote, her theme—that CBC Radio had grown moribund and must quickly adapt or die—was already resounding among radio management. My first inkling that revolution was once again in the air came in March of that year when Donna Logan, then the head of English radio, made a manifesto-like address to the CBC Radio Council—a now-defunct national consultative body. Reflecting on the end of communism, Logan predicted a period of "overwhelming change, change more dramatic and more radical than anything we have seen over the last few months ... change [that] will come with dizzying speed." The prospect she sketched was vertiginous—technical changes that had occurred in the last five years alone, she said, had already determined that "we are in a different business now that we were in 1985." The trouble was that "CBC Radio hadn't changed." Logan diagnosed what she called "a bunker mentality" and suggested that we had become so hidebound and preoccupied by the daily grind that she felt the need to confide "quite frankly" that she was "worried [that] we may have lost our capacity to change." All the keywords in her address—"dizzying," "overwhelming," and "accelerating"—were designed to produce the same impression. The ground was shifting under our feet, a flood was sweeping over us—without rapid adaptation and the flexibility of a contortionist, we were surely lost.

Much could be said about this panicked mood. Some of it was a mid-life crisis of the radio revolution and the large and now aging generation—my generation—that had made this revolution and was now beginning to wonder whether its populist and nationalist battle cries meant all that much to new generations, let alone to radio listeners born in Somalia, Guyana, or Gujarat. Some of it arose from a much more general disposition. Business writer Tom Peters, then a best-seller, wrote of "A World Turned Upside Down"—a world, he said, in which all stable points of orientation had been lost. One of his popular titles was *Thriving on Chaos*. "To meet the demands of the fast-changing competitive scene," he advised, "we must simply learn to love change as much as we have hated it in the past."[2] Another best-selling futurist, Alvin Toffler, was so exhilarated and unhinged by the impending Information Revolution that he proclaimed "the overthrow of matter."[3] There would be no place to stand because there would be nothing to stand on. Many other examples of this strange compounding of anxiety and euphoria could be given.

What interested me at the time, and interests me still, is the way the word "change" was used. As one can see in the quote above from Tom Peters, the word had no distinct referent—it did not describe some specific change that might be judged good or bad. Rather it was a command—an endless, unlimited imperative mood in which one was to live: *Change!* Willingness to change became a sign of the good corporate citizen, the team player—unwillingness diagnosed a foolish and unrealistic attachment to a world that was rapidly drowning. A taste for such all-seeing, all-purpose words was one of the products of the management training that CBC executives were by then expected to undergo. One frequent resort was the Niagara Institute, established in 1971 as the "leadership development division" of the Conference Board of Canada. It was, by its own account, a "trusted provider … of high-impact, high-quality leadership development solutions."[4] The essential idea behind all such training is that there is a science of management which is quite indifferent to the activity being managed. It didn't matter whether you were producing radio programs or fly-swatters, you could equally well be "equipped with a practical process … to lead employees … through any change initiative."[5] Managers trained in this way could manage anything, according to a set of general rules by which the people who worked for them could be organized, aligned, and made productive. The unlimited generality of *change* perfectly suited their universally applicable techniques.

The process presaged by Bronwyn Drainie's op-ed and instituted by Donna Logan's ode to change came to be called Creative Renewal. Its first practical manifestation was the appearance of a blue booklet entitled "Program Development Guidelines," which appeared one morning in mailboxes throughout the radio service. Alongside it sat a foam rubber likeness of a light bulb, marked Creative Renewal, and signifying, one presumed, the illumination we were about to receive from the little blue book. The book began with our corporate "vision," then proceeded to our "core values"—a kind of word salad of terms with expansive positive connotations like "passion," "integrity," "respect," etc. After that came "strategic imperatives," expressing our desire to be "flexible," "networked," and "innovative." Our "character, personality, and profile" were then sketched with another group of happy words—"nimble," "accessible," "trustworthy," and so forth. And, finally, our proper "style, tone, and manner" were specified—"agile" and "playful" but at the same time "reliable" and "thoughtful." A note on this page also lets us know that "by always drawing from [this] approved list of descriptors, we can be assured of a consistent, synergistic and strategically correct style, tone, and manner."

I'm sure that many more of my colleagues than I ever got to talk with found these guidelines somewhat ridiculous. Certainly, no one ever wrote a good radio program, then or now, by first consulting an "approved list of descriptors" or attempting to maintain a "strategically correct manner." The document, in that sense, parodied itself. And yet, such things have their effect, in the same way, perhaps, that advertising has its effect—you may not buy the particular truck whose praises are being sung but you remain immersed in the idea of buying a truck, or something else that will provide the same life-changing satisfaction. An atmosphere—a climate of opinion—a *zeitgeist*, to give it a fancier name—is being created and, as it becomes more prevalent, it gradually wears away resistance. At this elemental level, what was being said was, first, that radio production, like management, has a technique that applies in all circumstances. The subject of a broadcast doesn't dictate the style—the "strategic imperatives" of the corporation do. It says that the program-makers who are subject to these imperatives are, and should be, all more or less the same—interchangeable parts. They must bend to the "CBC Radio Character and Personality" rather than bringing their own character and personality to the job. Programs, likewise, must adopt the "approved ... tone," rather than cultivating their own individual approach—one could not easily imagine the idiosyncratic styles of CBC broadcasters like Bob Kerr or Allan McPhee, Clyde Gilmour, or Arthur Black, passing muster, though all four were still on the air at the time these directives were issued.

The "Program Development Guidelines," moreover, fit into a regime that had been developing for some time and was now coming into full flower. One of its elements was the science of demography. The guidelines announced at the outset that they were based on "two primary market facts"—the first that the current CBC audience was aging; the second that it was not being replaced "at an equal pace." "Even our most loyal listeners," the booklet averred, have begun to find us "boring and earnest." (No examples were given, nor were any programs mentioned.) The solution was to try to attract new listeners among people between the ages of thirty-five and sixty-four—this was to be our "target demographic."

By the early 1990s, such a phrase affrighted few ears at CBC Radio, but once it would have clanged intolerably, and perhaps even have seemed indecent.

In the first era of public broadcasting, one certainly expected farmers to listen to the farm broadcast, women to tune in *Trans-Canada Matinée*, and children to enjoy *Maggie Muggins*—my first love on CBC Radio—but there was no sense of targeting or preferring one segment of the population to another. One wanted listeners but accepted that they would select themselves according to their own liking—a "strategy" by which to seduce them, or a "target" at which to aim, would still have been unthinkable. The mode of address was primarily citizen to citizen—a public service engaging "the public"—rather than producer to consumer. The CBC still maintained a respectful distance from its audience—a reserve based on the recognition that "the listener" was a fellow citizen and compatriot but otherwise still an unknown stranger. Only gradually did the idea creep in that this listener could and should be subjected to a detailed anatomy, assigned a demographic box, and then tickled in exactly the place this research indicated.

It all felt, to me, like a panic, a manic flailing after "relevance" that was uninformed by any definite sense of vocation. We weren't asking what it was that we ought to do at this moment—we were asking how we could fit in better. This was evident in the language of the Program Development Guidelines—it was all about adjustment and image, strategy, and brand. We wanted to be more "with it" without ever asking what *it* was or where the pursuit of change for its own sake might lead. A new sensibility had begun to take root at CBC Radio.

Signs of the Times

One sign of the changing times was the condemnations one began to hear of "silos"—management-speak for any outpost of resistance to the new corporate culture. The word took its colour from the heavily fortified underground emplacements in which nuclear missiles are stored. Silos protected people who were not team players. Team players must be flexible and adaptable, sliding easily from niche to niche according to the corporation's ever-changing needs. Workspaces were redesigned to promote this new culture of collaboration. When the CBC's English network headquarters was consolidated in 1992, and people moved from workplaces scattered all over downtown Toronto into the new Broadcasting Centre, this process accelerated, and more and more people found themselves in open spaces, or open concepts, rather than offices. Soon we had a "Workplace Revitalization Project." After a period of consultation, it released a pamphlet which inveighed against "walled fortresses" and "inward looking ... gated communities," which "reinforce ... an insular culture of ...

hierarchy and division … status and seniority." It put forward a "vision" of "dynamic space with … permeable boundaries, encouraging flow and collaboration." Among its proposals was that "each team" should have "a sandbox"—"a dedicated floor area"—and "a kit of parts" by which that team could configure the space to its current liking, and then endlessly re-configure it.

Not all these dreams of plastic spaces for plastic people came to pass, but there were changes. The glass-walled meeting rooms that were created were one striking example. On their glass walls were embossed jumbled clouds of words in different sizes and fonts, creating an impression of typographical apocalypse—the age of print remembered and dismembered in the same gesture—and offering an apt sign of the haze of indefinite significance, or meaning without limits, that characterized this new order.

Another important innovation was our redefinition of ourselves as producers of "content." This began, in my recollection, with the creation, in Winnipeg, of something called "The Content Factory." One of its founders, Chris Boyce, went on to become the head of CBC Radio, until he was forced out in connection with the charges of sexual assault that were laid against his brightest star, Jian Ghomeshi, in 2014—a story I'll take up later. The name *Content Factory* had a cool, up-to-date sound, achieved by tying the new—everything is information—to the old—the industrial age image of the factory; but what it did was to deprive the craft of radio production of all specificity. "Content" describes the work of CBC Radio by a term so smooth, so general, and so inert as to be virtually featureless—it evokes a staple and, implicitly, interchangeable product. Whatever is broadcast—be it music or news, drama, or current affairs—is content. To speak of it under that heading is always to emphasize the most general aspect of what one is doing. All is grist for the great mills of Culture Inc. It's a demoralizing word, but it succeeds by claiming to be realistic—an adaptation to the new ways in which people are "consuming our content" and a recognition that that's finally all it is—a consumable and disposable product.

A telling example of this new outlook was the decision to discontinue the transcription of *Ideas* broadcasts. This was a service which we had provided to our listeners since 1982 and which I found invaluable in several ways. Transcripts could be stored, distributed, and reproduced in ways that audio, whether analog or digital, could not; they could be consulted and cited in ways that the audio could not. Moreover, they were a form in which the many scholars and commentators who made themselves freely available to *Ideas* producers could be acknowledged and recognized in the print milieu which, for most of them, was their primary theatre of operations. This was a vital form of reciprocity with the academic and cultural scene in which we worked and on which we depended. But around 2005 transcripts were suddenly found redundant by those holding the purse strings. We now had so many "platforms," it was said, why bother with an obsolete print record? The arguments I vainly attempted about the practical uses of such a record gained no purchase on those making the decisions. This was an instance, it seemed to me, of the generalizing and homogenizing habit of thought that is embodied in the word *content*. Reciprocity with sources, the balance between the spoken and the written word in *Ideas* production—these were considerations applying to actual radio programs produced in a definite and constraining context. In a world of *content* dispersed across multiple *platforms*, they sounded antique.

Another idea that was beginning to assert itself in the years leading up to Creative Renewal was "branding." No longer could an *Ideas* show run its full length without a pause or two to impress our brand into our listeners' minds. No longer could our networks do without slogans. These were sometimes rather lame slogans like the tags "News and More" and "Classics and Beyond" that were adopted when our networks were renamed Radio One and Radio Two in 1997, but the principle is what I want to emphasize here. Soon we would be regularly drilling into our listeners' minds the idea that "Canada Lives Here." Again, the CBC was only conforming to what the surrounding world presented as an iron necessity: brand or disappear below the horizon of popular attention. I have not forgotten the morning in the 1980s, when I was cycling through the University of Toronto along St. George Street and looked up to see that, on every lamp post, banners now proclaimed: "Great Minds for a Great Future." St. George Street had then just undergone an expensive makeover, compliments of architect and Bronfman scion Phyllis Lambert, and these pennants were evidently the finishing touch. Today the university sings its own praises on every street in its vicinity, but at that time it still shocked me that a university would shill for itself in this way. The CBC was obeying the same imperative. Modesty and restraint had ceased to be virtues and become signs of pride and affectation. The word *brand* no longer evoked the red-hot iron and seared flesh by which a rancher identified his stock. Branding, like "messaging" had become natural—the way in which we all must make ourselves known and leave our mark.

Words matter. They entrain thought and then fossilize it, as the word becomes familiar and the thought embodied in it invisible. This gradual desensitization brings to mind the old fable of the frog in a pot of water that is heated so gradually and imperceptibly that the frog never detects the change and so is eventually boiled alive. As science, it is spurious, though experimental verification was once or twice claimed in the nineteenth century;

however, it remains an apt metaphor. Language in which we are immersed changes our thinking without our noticing. What once shocked now barely registers. I made the point earlier with regard to "demographic targets"—a term now fully naturalized at CBC Radio. Both the image of taking aim, and the treatment of listeners as typical expressions of an objectified population are inimical to the spirit of public broadcasting, but habituation blunts the effect. A term like *Human Resources* loses its overtones of exploitation and fades into the background as HR. *Content* is a word and a concept of this kind. Like a super-highway which flattens and obliterates the landscape through which it runs, it does away with all distinction and peculiarity.

Revolution from Above

Creative Renewal, at CBC Radio, belonged to a style of management-led revolution that was then just beginning and has remained common since. Everywhere there was talk of re-engineering, reinvention, and, a little later, disruption. This was a revolution that had neither bubbled up from within, nor been imposed from without. It was designed in the management suites, like the "approved descriptors" and "strategically correct tone" of the Program Development Guidelines. Its first reason for being was the new cult of management itself. This was not the management about which producers had complained from time immemorial at the CBC. Management, according to this old complaint, was stodgy, timid, and bureaucratic—a dead hand inhibiting the creativity of its staff. Now the shoe was on the other foot. It was the staff, according to Donna Logan, who had "lost [the] capacity to change" and needed to be stirred up and rejuvenated by their alert, attuned, ahead-of-the-curve managers. Management was no longer a conservative vocation more or less tailored to the activity being managed. It was a universal and quasi-messianic technique for riding the waves of "change" and discerning the crucial "trends" that ought to orient and "position" CBC Radio. People schooled in "high-impact, high-quality leadership development solutions" needed to put these new skills to work.

The second thing driving the panic that was Creative Renewal was new knowledge tools. It was becoming clearer every day that networked computers were going to bring about a new age—an age in which many of the assumptions on which public broadcasting had so far rested might become obsolete. This prospect seemed to induce an almost universal giddiness, expressed by manic swings between euphoria and anxiety. The possibilities and the pitfalls both seemed limitless, and this certainly contributed to the idea that one must simply "change" without regard for what might be lost and without inhibiting judgments about where our new plasticity might take us.

The difficulty with all of this, as far as I was concerned, was that CBC Radio seemed to lose its moorings. The rhetoric in Donna Logan's "A Time for Change" was a farrago of demography, management science, futurology, and pop cultural tea leaf reading. It didn't reflect on what the CBC had been, or on what it ought to be, according to some philosophically grounded account of the purposes public broadcasting serves. It was all about adaptation, relevance, and trend-spotting. The mood, as I've said, was one of controlled panic—we were falling behind, we were old hat, and we were going to miss the boat. Everything depended on our keeping up with the fast-moving, ever-changing blob that some were beginning to call "the culture." There was no thought that our "mandate" might consist in the construction of a place to stand—a stable ground from which to analyze and understand the whirligig of apparent change.

The question, at root, is where CBC Radio gets its standards. Is its job merely to keep up, to cater to the changing tastes of its current demographic "target," to march to the drumbeat of constant change? Or should it cultivate more stable and enduring standards? I would argue that the old CBC—the CBC that Donna Logan thought in desperate need of "change"—still existed in a relatively stable framework. Often the CBC had been a leader and an innovator, but it also looked outside itself for guidance, its agenda modulated by Canada's other social, cultural, and political institutions. Prominent Canadian scholars, like George Grant or Northrop Frye, were frequently heard and seen on the CBC because the CBC was connected to the university milieu in which their stars had risen. The music presented on the CBC was the music Canada's most accomplished musicians in all genres recommended to it. There was a stabilizing reciprocity between the CBC and its cultural setting. This give-and-take disappears from Donna Logan's vision. She has the CBC pursuing evanescent "trends" rather than conversing with stable institutions. Adaptation, flexibility, and adjustment have become the only standards. The CBC's situation is imagined in such fluid and ahistorical terms that only a kind of feral journalistic instinct can keep it abreast and in tune.

In such a situation, the "leadership" supplied by management becomes crucial. The "knowledge workers" of the future, said Donna Logan, "would require leadership to bring out their best performances and to make them adapt and change quickly." The "program development guidelines" were, presumably, an example. We needed to think for ourselves—"agile," "confident," and "bold" were all on the approved list of descriptors—but we also needed consultants who could tell us how to maintain the "strategically correct style, tone, and manner" without having to refer to the subject or the occasion. Programs should "break new ground," but the new, freer, more creative radio service that would do the breaking was to be the outcome of a tightly controlled and pre-meditated process—a kind of planned spontaneity, or engineered surprise.

Behind Creative Renewal lay a remarkable confidence in both the prescience and the scope of management, and an equally striking disregard for how new and worthwhile things actually come about. Neither character, nor commitment, nor inspiration can be induced by "high-impact ... leadership ... solutions." Such excellences can, at most, be hoped for, awaited, and nurtured when they turn up. Real character is the very opposite of the ingratiating affability that is described in the "program development guidelines" under the heading "CBC Radio Character and Personality Profile." Character suffers itself as an inescapably individual destiny. It expresses itself as a unique voice. It cannot be conjured up by management consultants. No one could have known they were looking for Barbara Frum, Peter Gzowski, or Michael Enright before they turned up because their styles were entirely their own. And so it is with commitment. It was their deep and learned love of music that made broadcasters like Lister Sinclair, Bob Kerr, and Ken Winters such engaging guides to its enjoyment, not some pose assumed in relation to a demographic target. Inspiration, finally, is even less biddable. For talent to find its occasion and its voice, someone must recognize and foster it, but that is all that can be done. Inspiration, by its very nature, can be neither planned nor commanded—it can only be allowed.

Creative renewal supposed that teachable technique was what was going to matter as we entered "the nineties." (Donna Logan's "A Time for Change" was subtitled "CBC Radio in the Nineties.") There was no time for patient and attentive waiting—no time for deliberation—no time for the slow cultivation of distinctive voices. Indeed, there was barely any time at all in the breathless "adapt or die" world mapped out by Donna Logan and the management gurus on whose works she drew. Time was evaporating like smoke before our eyes. From now on we would have to run just to remain in place. A new template was being set for CBC Radio, and it's not hard to discern the future we were preparing for ourselves in it. CBC Radio would not lead; it would follow. It would not originate; it would adapt. Above all, it would be likeable—an engaging puppy ever ready to lick your face—and it would be correct—an enforcer of approved thought styles rather than a questioner of them.

Participation in this new account of public broadcasting was to be mandatory, as Donna Logan made clear in the ominous closing words of her address. Having established her point that the 1990s would be a time of almost unimaginably rapid and disorienting change, she then argued that only a "united radio service" would have any chance of successfully meeting this daunting challenge. "If we are going to thrive in the '90s," she said, "we have to agree on a vision and work on it together." There would be "extensive consultation" but, after that, our ranks must close. "Once we have agreed on what it is we want to do," she warned, "I expect people in the radio service to buy in." And then came the kicker: '*There will not be room for those who cannot.*'"

At the very meeting at which she spoke, I received confirmation that her threat might be more than a rhetorical flourish. I had risen to speak about the prevalence of "plastic words" in the discourse of Creative Renewal. This was a term I had recently learned from a professor of German literature at the University of Freiburg, Uwe Pörksen, who had published a book by that name in Germany in 1988. (I would later present his work on *Ideas* and participate in the English translation of his book, which the Penn State Press published in 1996 as *Plastic Words: The Tyranny of a Modular Language.*) A plastic word, according to Pörksen, is a word of immense and expanding positive connotation, and no denotation whatsoever—"it makes waves but doesn't hit anything" was Ivan Illich's pithy précis of the idea.[6] "Change" as Donna Logan used the term fits this definition precisely. No actual change is indicated just the vast, radiating positivity of endless changeability. Such a word possesses no ambiguity, evokes no history, catches no nuance, and casts no shadow. It is, in essence, a command—"a light that is never turned off," as Pörksen would later say in our interview for *Ideas*.

I unfolded this idea, as best I could, for the Radio Council, and then sat down. The next to rise was Alex Frame, then the second-in-command at CBC Radio, and later our commander-in-chief. "I agree with everything David Cayley just said," he told the meeting, "except for the career-damaging part." Alex Frame was a man I liked, and who, I believe, liked me, but his threat was, nonetheless, quite explicit. The point was reinforced, a few years later, when Frame, by then the head of radio, paid a courtesy call on the *Ideas* annual planning meetings. In the course of a group discussion with him, I asked if he saw any role for a "loyal opposition" to the current regime in radio. I chose my words carefully, and I believe he understood that I meant to give the term the full sense of a disagreement within an overarching agreement that it carries in parliamentary tradition. His answer, even so, was an unequivocal no. There was, as Donna Logan had said, "no room" within CBC Radio for a variety of interpretations of the purpose of public broadcasting. The talks department may once have resembled a faculty club, as Eric Koch, one of its members, used to like to joke in later years, and *Ideas*, in my time, may have been its last echo—"the academic senate of the CBC," in the affectionate caricature of one of our frequent contributors—but CBC Radio no longer bore much resemblance to a university and its producers had little in common with tenured professors. (The contemporary university didn't bear much resemblance to a university either by that point, but that's another story.) We were part of a strategically managed corporate team, and we needed to stay "on the same page"—one of a flock of new clichés like "getting it," staying "on message," etc., that emphasized the importance of remaining attuned to the evolving hive mind of "the culture."

Luckily, I never suffered the career damage at which Alex Frame hinted. In truth, I didn't really have a career in the sense that then prevailed at the CBC of a carefully curated progress reviewed annually with one's manager. (An annual conference with every employee to assess their "professional development" was mandated by the CBC's contract with the Canadian Media Guild.) *Ideas* was the only place I wanted to be, and the only place I could make the programs I wanted to make, so I wasn't going anywhere anyway. *Ideas* then had a peculiar status within CBC Radio as a show often praised but less frequently listened to. It was venerable: *The Best Ideas You'll Hear Tonight* went on the air in 1965, but it descended from two earlier shows, *University of the Air* and *The Learning Stage*, that went back to the 1950s. And it was believed to be serious, which made it a strategic asset that could be pointed to with pride when the CBC's "mandate" was in question. But, beyond that, I think CBC Radio management generally saw the show as something that was good for others, rather than as something of interest to them. (Harold Redekkop, who served as head of CBC Radio between Donna Logan and Alex Frame was a happy exception.) This meant that it was probably safe enough for me to inveigh against plastic words at the Radio Council, to defend craft traditions and habits of thought peculiar to the *Ideas* "silo," and to reject the idea that it was healthy for CBC Radio to speak in a unanimous voice. But what were the implications of this new climate of opinion for CBC Radio as a whole?

Implications of Creative Renewal

I think that there were four main consequences. The first was that CBC Radio became painfully self-conscious. Our vision, our tone, our manner, our "brand personality"—all could be designed, managed, and "bench-marked," as the current management lingo had it. The implication was that we would spend a lot of time looking in the mirror, fixing our hair, and adjusting our smiles. Rather than judging our adequacy in relation to our times—how well were we meeting the challenge of representing and analyzing the world around us—we would judge in terms of whether we were meeting our internally generated corporate standards. Rather than submitting to the ordeal of our times, we would scrutinize our own style—were we sprightly enough? Had we overcome the deadly "earnestness" that the program development guidelines argued was oppressing even "our most loyal listeners"? Were we "breaking new ground"? Did we, in short, have a winning personality?

The second consequence was that we began to lose our connection with our own history. To wish, at all costs, to change, and to argue that this imperative operates regardless of all substantive considerations is to lose one's roots in the past. I do not speak of some idealized past. The CBC's history, as I have shown, has been a continuous compromise with a frosty political milieu, a hostile private broadcasting industry, and a public whose tastes have already been shaped by the blandishments of Hollywood and New York. I'm referring only to the memory of an experience that we can call our own—a definite character, rather than a plastic one capable of being remoulded on command. In the world of plastic words, history becomes a laboratory, in which reality is manufactured, rather than a set of constraints defining a field of action.

The third change was the crossing of the threshold between reality and representation. We were entering the domain of what French thinker Jean Baudrillard called "the simulacrum," and his contemporary Guy Debord "the society of the spectacle."[7] The great issue, in the discourse of Creative Renewal, was not our fidelity to our place, our history, or our legislated purpose, but whether we were sufficiently in tune with the looking-glass world of media.

The final change was that a social and political orthodoxy began to form. According to the "boring and too earnest" stereotype haunting the program development guidelines, the old CBC was a tiresome pedagogue, whereas the hip new CBC would be a cool, stylish non-judgmental friend. That was the theory: The CBC would broaden its range in the new regime. What happened in fact was that CBC Radio began to narrow its scope. The old CBC had conceived itself as a forum—a place where different opinions engaged with one another. The idea was embodied in first-era shows like *Citizen's Forum*—the discussion program that ran for twenty-two seasons on CBC Radio between 1943 and 1965—but it continued in many post-radio revolution programs like *Between Ourselves* or *This Country in the Morning* and its successors. The idea, at its best, was to make the country known to itself as it was, rather than to model how it ought to be. The new mood lent itself much more to what was then called political correctness. Part of being in tune and in touch lay in holding and expressing the right views. The range of political and philosophical opinions heard on the CBC began steadily to shrink, until it finally reached today's quite remarkable uniformity.

Conclusion

Creative Renewal was intended, as its name says, to renew and refresh an existing inspiration. Alex Frame wanted to reawaken an old spirit—by "shak[ing] up radio."[8] Bronwyn Drainie hoped to relight an old fire by her call for revolution. I have tended to emphasize, on the other hand, what was novel in the spirit of Creative Renewal. Both perspectives have their point. Creative Renewal *was* a dramatic departure from previous CBC Radio norms—in its managerialism, in its acute self-consciousness, and in its incipient orthodoxy. But it was also a continuation of the populism that I described earlier. This was true, above all, in its rhetoric of change and adaptation. The call to get *with it* supposed that there was still an *it* to get *with*. When people spoke of *the* culture, as they increasingly did, they still indicated a singular entity that CBC Radio was supposed to serve and from which it could draw its standards. The public might be undergoing continuous transformation, as it seemed in Donna Logan's speech, but one voice still spoke from the whirlwind of change. Producers might have to be more nimble and assemble their brand personality from the "approved list of descriptors," but with the right leadership, they could still learn to discern what ought to be done. The single people, from whom populism derives its mandate and takes its instructions, was still there and still issuing its commands.

5 | The Ghomeshi Effect

Creative Renewal initiated populism's final phase at CBC Radio. But revolutions, even minor revolutions, don't complete themselves until all the veterans of the old order have finally left the field. Consequently, Donna Logan's call for unending change, and the ingratiating brand personality recommended by the Program Development Guidelines, took effect only gradually. Bob Kerr, whose old-fashioned courtesies and unassuming erudition epitomized a much older CBC, continued to host *Off the Record* until his retirement in 1996. Peter Gzowski made his final *Morningside* broadcast in 1997. I lasted on *Ideas* until 2012, and Michael Enright, the last of the old school, didn't retire from *The Sunday Edition* until 2020. So, for the time being, the fact that Alex Frame didn't want, and wouldn't recognize, a loyal opposition didn't mean that I lacked sympathetic interlocutors with whom to mutter darkly over lunch.

However, change did come, and its preeminent sign for me, was the arrival on CBC Radio of '*Q*' hosted by Jian Ghomeshi in 2007. Today, Ghomeshi is mainly remembered for the accusations of sexual assault that led to the CBC's dismissing and disowning him in 2014. He has remained a non-person since, despite his acquittal at trial in 2016. But, to me, his sexual habits, whatever they may have been, are overshadowed by the change he represented, and helped bring about, in the broadcast style of CBC Radio. '*Q*'s hallmark was total immersion in popular culture. This preoccupation had many antecedents at CBC Radio, notably *Prime Time*, which ran from 1987 to 1993 with Ralph Benmergui and Geoff Pevere as the main hosts, and *Definitely Not the Opera* (*DNTO*), with Nora Young and Sook-Yin Lee, which lasted from 1994 to 2016. But '*Q*' still marked a departure.

What struck me first was Jian Ghomeshi's evident, if unarticulated, assumption that popular culture constituted a kind of second nature for his listeners—a taken-for-granted environment, rather than a suite of commercial products that would be known to some of his audience and not to others. In writing for *Ideas*, my colleagues and I were generally scrupulous about blind references that might make listeners feel they had strayed into a club to which they did not belong. It might be unnecessarily pedantic, for example, to qualify G. W. F. Hegel as an "early nineteenth-century German philosopher," but it was preferable to assuming familiarity by a casual reference that might make a listener feel that anyone who had not read Hegel's *The Phenomenology of Spirit* might just as well turn off their radio right now. On '*Q*,' on the other hand, an encyclopedic knowledge of popular television, movies, music, and celebrity gossip was simply taken for granted. For example, I remember one morning being mystified by a reference to "hair bands"—bands of the early 1980s, as I later discovered, whose members typically had big hair—but it could just as easily have been a learned discussion about a TV series I had never seen, or a careful parsing of autobiographical overtones in a song I had never heard. Intellectual culture needed signposts; popular culture was supposed to be a home ground on which every outcropping was known.

But popular culture isn't quite the right word, coming, as it does, from a time when there was still a distinct high culture, and the remnants of a folk culture, and a working-class culture, with which to contrast it. After the 1960s, these distinctions were progressively erased. In the new departments of "cultural studies" that began to be established in universities, one could as plausibly study Batman comics as *Finnegan's Wake*. The latter might be harder to read, but they both belonged to the same flat landscape. When the head of the Ontario Arts Council told philosopher George Grant, in the 1960s, that Mozart's *Marriage of Figaro* was "just the *South Pacific* of its age," Grant was sufficiently shocked and appalled that he recalled the remark in a conversation with me twenty years later.[1] Today, Grant would have a hard time convincing anyone that Mozart stands on a different plane than Richard Rogers. One might express a preference, but Grant's assumption that Mozart demands and deserves attention of a higher quality than Rodgers and Hammerstein's hit musicals belongs to the past.

Popular culture had become simply "the culture"—a term I heard repeatedly used on '*Q*' to signify a cultural ecology that was treated as something as natural as a forest, a meadow, or a wetland. This culture constantly evolved and changed, throwing up new forms, outmoding others. It was conceived as a living organism, which, of course, culture had always been, but this was the first time that commercial culture, as an ensemble of engineered devices, designed to amuse, had ever been naturalized in this way. Aesthetic judgment flourished—one talked endlessly about what was in and what was out, what was generating buzz, and what was becoming a drag—but moral judgment was *verboten*. The aesthetic of the hit, a taste for the cool, attunement to those little bursts of glory that radiate from what is perfect of the moment—these governed the show's sensibility. A sense of constant excitement was generated—the awesome, the amazing, the unbelievable were always at hand.

Jian Ghomeshi was a paragon of this sensibility—an impresario of cool—a celebrity among celebrities able to precisely gauge what mixture of deference and presumption each guest, artist, or commentator was due. In its second season '*Q*' moved from the afternoon to a more conspicuous mid-morning slot and enjoyed immediate success. After *Morningside* had ended with Peter Gzowski's retirement in 1997, the CBC had tried in various ways to replace it with similar programs—first, *This Morning*, and, then, *Sounds Like Canada*. But both suffered from the attempt to preserve and extend a formula that had depended heavily on Gzowski's bashful charm, rather than developing a character of their own. '*Q*' was a new departure, and, in time, it developed ratings that exceeded Gzowski's, as well as being picked up by over fifty American public radio stations. CBC Radio had rarely had such a success.

An Inkling of Glory

A decisive moment in my understanding of the sensibility of the new program occurred in the spring of 2009. The occasion was an interview with Leonard Cohen that Ghomeshi recorded at Cohen's home in Montreal. It was in listening to this interview that I learned to connect the contemporary use of the word *icon* with its traditional use. The Christian church gave this word its canonical definition at a council held at Nicaea in the year 787. An *icon*, the council said, is "a window into eternity," "a threshold at which [an] artist prayerfully leaves some inkling of the glory which he has seen [beyond] that threshold."[2] Today, the word is ubiquitous and can be applied to whatever has entered the heaven of recognizable brands, whether person, product, or design. The fact that these definitions remain connected became clear to me while reflecting on the way in which Ghomeshi treated Cohen.

During the interview, Cohen made several unsuccessful attempts to introduce the name of Montreal poet Irving Layton. Layton had been a friend and poetic mentor during Cohen's youth, and it seemed that Cohen's return to Montreal was bringing those days back to mind. As an interviewer myself, I thought, *he's told you what he's interested in talking about—follow the lead he's given you*. But Ghomeshi passed over these invitations and kept the conversation on the track that he and his colleagues had presumably planned for it. Cohen was to be frog-marched through the various stages of his ascent to iconicity. Ghomeshi used the term repeatedly, as if to remind Cohen of his status. This was not just an encounter with a writer, singer, and eminent Canadian, I realized—it was an access to divinity.

Today, it's a rare episode of '*Q*' in which some *icon* is not presented, remembered, or expected. It could be argued, of course, that the proximate source of the term is the computer icon—the clickable logo that opens into cyberspace—not the object of religious adoration. If this were so, it would still be evocative enough. The computer's infinite inside/outside, everywhere/nowhere, is already a kind of immanent heaven, which absorbs many of the former functions of religion. But I think it's the term's Christian roots that are more clearly on display in the way in which Jian Ghomeshi approached Leonard Cohen. What I felt I heard, listening to Ghomeshi interview Cohen, was that Cohen was not quite a mortal. He was not just a man in Montreal whimsically recalling an old friend. He was a threshold at which Ghomeshi could gain access to a certain blessedness and, while basking in this supernal light, also share its glow with his listeners.

An icon, of old, was an "inkling" of the divine glory. What now makes Leonard Cohen one? The man himself did not appear avid for such a status, at least on the day in question. He was evidently a bit bored with Ghomeshi's recitation of the litany of his achievements and eager, if possible, to introduce some more congenial subject. Yet Ghomeshi felt he ought to persist—presumably on the grounds that icons must be treated as icons. The first component of this new dignity, I would say, is sheer celebrity—a word that already puts us on holy ground, if we take seriously its derivation from the old French *celebrité* for a "solemn rite or ceremony." The second is the romantic myth of genius in which various ideas of divine possession are compounded. (This, obviously, does not apply to a Coke bottle or a 1955 Chevrolet Bel Air—both undeniably *iconic* in the sense we are exploring—but is apt for Cohen.) Together they create a compact quality which is present and available in the icon and able to induce a state of grace in those who come in contact with him or her. Ghomeshi, as I came to understand him, was a curator of this quality—and, at times, its priest. To some, he may have seemed vain and self-satisfied—and he was achingly, endlessly cool—but he also possessed the humility this curatorial task demanded—he wasn't going to treat an icon as just another man—that would threaten the whole carefully graded hierarchy to which he commanded access.

'*Q*' achieved, as I've said, unprecedented popularity in its time slot, as well as wide distribution in the United States. Before this time a few American public radio stations had carried *Ideas* or *As It Happens* but not on anything like the scale achieved by '*Q*.' This success made '*Q*' something of an archetype of the new CBC Radio. It set a tone that was adopted by many CBC

Radio programs. The most outstanding feature of this new tone, for me, was the way in which it erased what little remained of the foundational modern distinction between the public and the private.

Radio had been, from its origins, an invasion of privacy. Voices from afar were suddenly, intimately there in the midst of the quintessentially private life of the home. This privilege was usually acknowledged during the early years of radio and television, and both, for a time, tried to behave as well-mannered guests. A certain formal courtesy was observed, as well as a measure of recognition of the actual distance between the apparently present voice and its faraway source. But a long history now separates these antique courtesies from the confiding, just-between-ourselves tone that is now common on CBC Radio, and many stages mark this way. My generation, which came into radio in the 1960s and 1970s, already sought a much less formal relationship with the audience than the one cultivated by the old CBC announcer—a man, and now and then a woman, whom one could still imagine in the evening dress required of the BBC announcer in the early days of radio in Britain. CBC broadcaster Max Ferguson parodied the type in the comic figure of the orotund Marvin Mellowbell. Janet Somerville, one of the first producers of *Ideas*, remembered the announcer's virtues affectionately in a tribute to *Ideas*'s first host, Ken Haslam. Haslam, she recalled, "read what we wrote in his rich, beautiful, decorous voice evening after evening, patiently, courteously and with a touch of solemnity."[3]

But the style Somerville praises was beginning to lose its fit even as Haslam was adding gravity to *Ideas*. Barbara Frum's disaffection with "mid-Atlantic accents" is pertinent. One of the exemplars of the new style was Peter Gzowski. Gzowski as I argued earlier, sought a much more direct connection with his listeners than was compatible with the "decorum," in Somerville's nicely judged word, that was observed by Ken Haslam. When *Morningside* made its last broadcast in 1997, the testimony I remember hearing most frequently from grieving listeners was that they felt they were losing a friend. But, for all the intimacy Gzowski achieved with listeners, a certain rhetorical distance still remained. Gzowski's Canada, as many have noticed, was an imaginative construction, and the art required to project and sustain this vision was audible to a careful listener. One knew, however dimly, that something was being made and conveyed across a vast space. Indeed, the evocation of the country's variety and extent were a great part of the show's appeal.

This was no longer true with '*Q*.' Gzowski may have been partly creating the country to which he was broadcasting, but he still projected himself outward into a world apart from him. Ghomeshi pointed inward—into the great maw of that second nature that he called "the culture." But this was not inwardness in the old sense of interiority. Rather it was a journey into an inside that had lost its outside. The historically determined Canada over which Gzowski's poetic imagination brooded was still a definite place, however mythically enhanced. Ghomeshi's world was pure *myth.* I use the word in the sense developed by French thinker Roland Barthes in his *Mythologies*, where he claims that myth is a device for removing history from language.[4] German scholar Uwe Pörksen, likewise, regards this as an outstanding feature of what he calls plastic words. They "grasp … history as nature," Pörksen writes.[5] I see Ghomeshi as having been engaged in this kind of naturalization. This does not mean that "the culture" was romanticized—it was often described in precise analytical detail, just as a naturalist might describe a lake or a stream. Nor does it mean that dissent was absent. In fact, a certain dissidence, heavily stylized, was *de rigueur*—the "rebel sell," as writers Andrew Potter and Joseph Heath call it, had been identified by journals like *The Baffler* and *Adbusters* when Jian Ghomeshi was still in short pants.[6] Nor, finally, does it mean that dreams of a transformed or perfected world were lacking—imminent transformation is bread and butter to this worldview. All I'm pointing to is the disappearance of a distinct reality standing apart from and constraining its representations. Ghomeshi was at home in what French philosopher Jean Baudrillard calls "hyper-reality"—a place, but not a place, where signs point to other signs rather than to some definable object; where models absorb the world they ostensibly describe; where events are so quickly assimilated to some already available significance that they seem to declare their meaning almost before they occur; and where people appear trailing so many demographic and diagnostic signifiers that they are, in effect, known in advance.

Across the Threshold

In the world that I'm trying to describe a certain threshold has been exceeded. Carl Jung calls this moment of excess *enantiodromia*, a Greek word he adapted to describe the moment at which a tendency, pushed beyond a certain point, turns into its opposite. Media, up to a point, bring the world before us in a new way. It's a cliché but probably still true to say that distaste for the Vietnam War was driven by the novel experience of "seeing" war in real time. Television made the war more *real.* But, beyond this point, media begin to replace the world rather than represent it. In Baudrillard's terms, the map which once described the territory begins to engender the territory. Marshall McLuhan called it "a world gone through the looking glass in which the human observer creates his own space and his own time."[7] My contention is that 'Q' and many other programs that followed come from this space in which mediated experience has overtaken and engulfed vernacular experience. I recognize that experience has been, at all times, mediated—preeminently by that *ur*-medium language—but I'm trying to draw a rough and ready distinction here between the hall of mirrors in which ironic self-reference prevails and what is left of a homemade world in which people remain in some sense *in touch* with their place, their words, and their work. 'Q' was born after the flood—the flood in which Donna Logan was already afraid we would all drown in 1990—and its genius was to be at home and at ease in this new state of super-saturation.

I spoke earlier of the shock I felt when I realized that Jian Ghomeshi did not have to explain his references, whereas I did. I was still engaged in a formal style of address—a style as solemn and decorous, in its way, as Janet Somerville had found Ken Haslam to be back in the early days of *Ideas*. Jian Ghomeshi addressed a "we" about whom he seemed able to assume almost everything. This, of course, did not begin with him in some absolute sense. There had been other CBC hosts who spoke as if their listeners were just like them. I can remember the dismay I felt one morning in 1991, when broadcaster Bill Cameron, who was filling in for Peter Gzowski on *Morningside*, began that day's broadcast by making a very big deal of the fact that it was Bob Dylan's fiftieth birthday. His mock incredulity was painful. Had he thought Bob Dylan immortal? Did he think that listeners who were eighty, or twenty, would care? It was the sort of self-centred tactlessness that got the boomers a bad name. And yet I still think that this already existing familiarity with the audience took a new and very long step with the style of programming of which I have been making 'Q' the emblem.

Listeners became "guys." The "how you guys doin'?" that you might hear from a concert stage, or a waiter in a restaurant, now came from CBC Radio as well. The assumption was that the listeners were all of one style, one taste, and one experience—they all knew what a "hair band" was/is; they all consumed the same movies, music, and television; they were all gripped by even the most banal details of the lives of those who created these entertainments; and they all shared the same pop music pantheon and assigned its upper echelon the same semi-divine status that Ghomeshi accorded Leonard Cohen. This cannot actually have been true of an audience which must have included some at least who had previously found in the CBC a refuge from many of these things. But that never seemed to matter much—these outliers and old-timers weren't in our "target demographic." CBC Radio had ceased to be recognizably public in any sense that was distinguishable from private. It spoke of private taste, of private feelings, of private associations. Gossip ceased to be gossip because there was no remaining privacy it could violate—it became news.

Charles Taylor, in his *A Secular Age*, characterizes modernity by what he calls "direct-access." In older forms of society, distinct ways of life, along with the many delays, detours, and intermediations to which central powers were subject, created protective local envelopes. In the direct access society, everyone stands on the same level; each one is immediately connected to the whole. There is no shelter. Courtesies, formalities, modesties, and decorum: all indicate the sense of a difference and a distance. They say: I don't know who you are or what you think. And, because I do not have direct access to you, I must precede by indirection, seeking a third space, a public space, in which we can face one another without dissolving our differences. When these distancing devices disappear, I can only lose myself in the whole, consenting to be the one whom the familiar voice from the radio assumes I am. And, with all difference gone, only the artifacts of commercial culture remain to unite us and provide a common ground, which is not a ground but a platform.

CBC Radio developed a new mode of address that assumed an identity between the broadcaster and the listener. There was an increase in the promiscuous use of the word *we*. If each one occupied the same space—the planet, as one casually said—the same time—the global simultaneity of the news—and had the same interests—the blockbuster movie, the hit song, the big game—then it was easy to speak of a great "we" as the collective subject of a common experience. Discourse was personalized, by design and by edict. The word "you" was used as frequently as possible to make a mass audience feel as if they were being personally addressed. Hosts took on the priestly role I earlier ascribed to Ghomeshi, pronouncing frequent benedictions: be safe, take care, have a *great* day, and so on. It seemed a long time since Bob Kerr had fondly wished his listeners nothing more than "a good afternoon."

All these gestures—of blessing, of personalization, etc.—work to collapse the distance between the CBC and its listeners. The entire production process is erased, and, along with it, all the many exigencies that govern this process—from the vast and expensive technological infrastructure to the journalistic conventions that determine what can and should be said. Everything must appear natural and unaffected, as if the voice emanating from the radio atop the refrigerator were really that of a friend whom I might later meet for dinner. Now, I know that if any medium were entirely given over to critical awareness of the many *mediations* it is performing, it would quickly succumb to a paralyzing self-consciousness and become too inhibited to talk at all. So, I am speaking, obviously, of a question of degree—of a critical point along a continuum running from formality to *naturalness*. The formality that the CBC once observed could sometimes descend into stiffness and pomposity, but it at least spoke implicitly of all that intervened between the broadcaster and the listener—its decorum, to continue with Janet Somerville's word, acknowledged a limit, and an unbridgeable difference. It is my contention here that the CBC has now gone very far—too far—in the opposite direction. A crucial threshold has been passed.

The way to this threshold, as I've tried to show, was long prepared. From the time of *Seven Days* on, the CBC broadcasters who were seen to be at the leading edge sought to become the delegates or agents of their audiences, rather than their interlocutors. The conviction that "the personal is political," which animated so many of the social movements contemporaneous with the radio revolution, also played its part in breaking down broadcasting's "fourth wall," as theatre people call the partition between actor and audience. At the CBC, the transition to a more immediate, more engaged, more personal style of broadcasting generated great excitement—in the exemplary case of *Seven Days*, when poor old Alphonse Ouimet swung in effigy from a gibbet outside the Vancouver Courthouse, this excitement became ecstatic, as it did among those who wrote love letters to Peter Gzowski. But pendulums rarely stop swinging in the middle of their arc, and clubbiness has long since replaced fustiness as the CBC's besetting vice.

Social critic Ivan Illich, in his writings of the 1970s, put forward the idea of "paradoxical counterproductivity"—his name for the point at which there is too much of some staple, like education or health care, and it begins to get in its own way and undermine its own ostensible purposes.[8] Jung's concept of *enantiodromia*, which I cited earlier, is a different but related version of the same idea. Any tendency pushed too far, says Jung, will reverse itself and become its opposite."[9] To put this in the terms I have been developing: excessive familiarity will finally increase the broadcaster's distance from the listener rather than reducing it; masking the artificiality of broadcasting's discourses by an elaborate show of naturalness will produce more, not less artificiality, and so on. The populism that intended to bring the CBC closer to its listeners now alienates those many listeners who do not share its premises. The promise to right "public wrongs" that brightened the *Seven Days* manifesto has petrified into sanctimony and self-righteousness. The ambition to "prepare Canadians for social change," which the Meggs Ward report felt should animate CBC Radio has turned into a narrow political correctness that preempts critical reflection. Concern has become sentimentality; immediacy has become a denial of mediation; advocacy has become an ideology.

Creative Renewal was supposed to refresh the radio revolution. Bronwyn Drainie, as we have seen, was explicit in her call for a "second revolution": Alex

Frame likewise wanted to "shake" CBC Radio out of what he feared was a fatal complacency. Revolution, understood as endless adaptive change, was the template. No one really knew what this intoxicating rhetoric of permanent change actually meant. Today, it's easier to see where it was leading. The CBC has lost any vital connection with its past; it has lost contact with a large number of Canadians who don't share its increasingly homogeneous worldview; and it has largely stopped trying to foster critical thinking at the very moment when nothing could be more needful.

Such total characterizations are chancy, I know, and overlook significant differences. Eleanor Wachtel of *Writers and Company*, until her retirement in 2023, interviewed with the same tact and careful preparation that had marked her approach since her show began more than thirty years before. *Ideas* still presents programs on Michaelangelo's poetry, the civilization of the Maya, and whether dogs feel guilt—to take a few recent topics. And yet it seems to me, nonetheless, that most of the people who produce and present the current crop of CBC Radio programs share the same politics, the same sensibility, and the same understanding of the CBC's vocation. This doesn't mean that I don't find some of these people charming and likeable, or that I am uninterested in the stories they tell. It just means that I seem to hear only one note played over and over again. Feeling predominates over thinking; story predominates over analysis; uplift predominates over understanding. I often feel I am in the presence of the votaries of a new religion who can't recognize their own religiosity, even though they are quick to recognize and reject rival cults. Populism has reached its paradoxical terminus in a new elitism.

6 | Television and the Politics of Canadian Broadcasting

In the previous three chapters, I have concentrated on the unfolding of the populist era in that part of the CBC that I know best—the radio service. But, of course, it is television that defines the CBC for many people, and television that has epitomized the fading fortunes of the CBC during the period I've been discussing. So, in this chapter, I will sketch the history of television from 1958, when the CBC lost its central position in Canadian broadcasting, to the first decades of the twenty-first century when Richard Stursberg set the direction in which CBC Television is moving today.

The CBC, from its beginnings, was a nationalist project. Graham Spry, Allen Plaunt, and their friends founded the Radio League in the conviction that they were answering the call of what Spry had called "the destiny that is Canada."[1] The agrarian crusade that Plaunt helped to spark in rural Ontario during the 1930s was named the New Canada Movement.[2] At first, it appeared that they had built solid and broad-based support for a predominantly public broadcasting system. Spry saw evidence of "a very profound movement of opinion" in the backing he gained from trade unions, farm federations, churches, newspapers, and politicians.[3] But this seemingly mighty coalition proved evanescent.

There was a weakening of the Empire nationalism that had led men like R. B. Bennett and Sir John Aird to see public broadcasting as a defence of the British connection and a bulwark against Americanization. By 1940, when Prime Minister Mackenzie King and US President Franklin Roosevelt signed the historic Ogdensburg Agreement, which effectively put Canada under American military protection, it was clear that King's long-lived Liberal government was gradually turning away from Britain and toward the United States. The newspapers ceased to fear private radio as a competitor and began to see the station owners as fellows in free enterprise. The Conservative

Party, resentful at its long exile from office, made the CBC its enemy and the Canadian Association of Broadcasters its friend. The members of the popular organizations that the League had mobilized proved more fickle as audiences, and more fond of American entertainments, than their political stance might at first have suggested.

Graham Spry said of R. B. Bennett that there was "a conflict in his soul," and this conflict, as I remarked earlier, was not just in Bennett's soul—it was in Canada's. Canada has possessed what writer Herschel Hardin calls "a genius for public enterprise," a genius, he says, akin to the American "genius for private enterprise."[4] This has shown itself in "crown corporations" that have ranged from the Board of Works, established to build canals in the Province of Canada in 1841, to Trans-Canada Airways, which was created a year after the CBC. The CBC has had many siblings, and necessity has usually been the mother of their invention, as it was with the CBC. Again and again, the choice has been the one Graham Spry set before the House Broadcasting Committee in 1932—the state or the United States. But, according to Hardin in his book *A Nation Unaware*, Canada has never properly recognized its own genius for economic improvisation. Instead, it has generally accepted the view of free-market liberalism that public enterprises are abnormal and ought to be abandoned or pushed aside, just as soon as some private alternative presents itself. For this reason, Hardin calls free-enterprise liberalism "the American ideology in Canada." It has used the glamorous promise of freedom to seduce us into the American market.

In the CBC's case, this ambivalence played out in the years between 1936 and 1958. In the 1936 Broadcasting Act, Canada clearly and explicitly created a broadcasting system that was to be regulated in the public interest by the CBC. But every time the CBC seriously attempted such regulation, an outcry from the private stations forced it to relent. This lax regulation was criticized by the influential Massey Royal Commission on the Arts in its report in 1951. "Far too many stations, regulated in principle by the CBC, offer programmes which must be described as regrettable," was the commission's haughty judgment.[5] But the fact was that the CBC had neither the popular nor the political support to actually exercise its formal power as the coordinator and cornerstone of a hypothetically unified broadcasting system. Looking back ruefully, at the end of his life, Graham Spry put it bluntly, "The CBC has never been created as provided by law and its powers have been progressively … reduced … or impeded."[6]

In place of the publicly regulated "single system," which the law mandated, a charade was created. Its first scene came to an end with the Broadcasting Act of 1958. The new Conservative government of John Diefenbaker stripped the

CBC of its regulatory authority and invested it in a new body called the Board of Broadcast Governors (BBG). The BBG, as I indicated earlier, then set about licencing the private television stations that formed the CTV network in 1961. This put the CBC in two binds simultaneously. The first was that it was now required to define and shape public broadcasting in a milieu in which private broadcasting would provide the template and set the tone. The second was that authority over broadcasting was divided, and the CBC was nominally made subject to the BBG as the source of its licence to broadcast. It was no longer clear who or what was in charge. Did the CBC answer to the BBG or its own legislative mandate? Was it a creature of the government that still held the purse strings, or the servant of its audiences as some of its producers were beginning to argue? These questions created confusions and deceptions that took place over several generations. I cannot tell the whole story here, but a few crucial episodes will give the idea.

The first notable crisis played out in the wake of the cancellation of *This Hour Has Seven Days* in the spring of 1966. *Seven Days*, as we have seen, was wildly popular, but it tore the CBC apart internally. The issue was the proper character of public broadcasting. Should the CBC be considered a public utility, operated as a public trust by a class of conservative "guardians," or an agent of social change led by crusading journalists whose mandate derived from their popularity?[7] Before the contretemps had run its course it became the subject of extensive hearings by the House of Commons Broadcasting Committee, a public inquiry, and even an intervention by the prime minister himself to keep the show on the air until the end of its second season. (One of *Seven Day*'s last irresistible bits of cheek was the opening announcement of its final broadcast; "Ladies and Gentlemen, by the good offices of the prime minister of Canada, *This Hour Has Seven Days* ....") Douglas Leiterman held a cloak and dagger meeting with Secretary of State Judy Lamarsh, the minister then responsible for the CBC, in the minister's limousine from which Leiterman then discreetly transferred into another car.[8] Patrick Watson lobbied openly in Ottawa to be made president of the CBC himself—an act of hubris which the gods didn't fail to punish when he was later made chair of the CBC Board in 1989 and found himself unable to accomplish any of his aims.[9] At the end of 1967, Judy Lamarsh was interviewed on television by Pierre Berton and declared "quite frankly" that she thought there was "some rotten management in many places in the CBC."[10] By then a new Broadcasting Act had already been tabled in the House of Commons. It became law in early 1968.

The Birth of the CRTC

The new act replaced the Board of Broadcast Governors with the Canadian Radio-Television Commission (CRTC).[11] Prime Minister Lester Pearson professed himself well-pleased. The "single system" had been re-affirmed, and the [CRTC] would prove itself, he predicted, "an effective piece of machinery."[12] But, according to broadcasting historian Frank Peers, the new act was no more than "an improved version of the act passed in 1958."[13] The fundamental contradiction that Pearson himself had shrewdly identified in parliamentary debate on the 1958 act was still in place.[14] The politic fiction of the "single system," as Pearson had said then, was, in fact, a smokescreen hiding the reality of a "dual system" whose components were at odds with each other. In 1958, Pearson had predicted that the new BBG would "tend to become a regulatory body for private stations only." He had also foreseen that the new board would be subject to regulatory capture because it would inevitably "be influenced … by the financial situation of [the] private stations." This was exactly how things had played out, but, ten years later, Pearson was no longer in a position to say so. Neither, it seems, was he in a position to fundamentally change the situation.

The *appearance* of change, however, was actively fostered. The "rotten management" issue was addressed by installing two proven managers. The prime minister's friend George Davidson moved over from the Treasury Board to replace Alphonse Ouimet as president, and management expert Laurent Picard became his executive vice president and president-in-waiting, taking over in 1972. Picard had been a research associate at the Harvard Business School in the early 1960s and had then become the associate director of the École de Haute Études Commerciales at the Université de Montréal. The new CRTC was also given an air of dynamism and cultural ambition. Pierre Juneau, a holdover from the BBG, was made the chair; CBC veteran Harry Boyle became the vice chair. The commission acquired an intellectual panache that had been mostly lacking at the BBG. Various other luminaries were added, including one of Canada's brightest academic lights, literary critic Northrop Frye. The job proved a trial for the dutiful scholar—he spoke to me, in an interview recorded near the end of his life, of "nine bloody years on the CRTC"[15]—but his presence certainly added to the impression that a new day had dawned, and a new broom was sweeping away the dreary remains of the BBG.

A brief honeymoon followed. In 1970, the commission held Canadian content hearings and passed the relatively strict regulations for which Juneau is still remembered at the Junos, the annual music awards that were named after him by grateful Canadian musicians. Both Juneau and Boyle became "nationalist heroes," and hope grew that they would call the CBC back to its proper vocation as a public broadcaster, when its licence came up for renewal in 1974.[16] Intervenors came from all over the country, at CRTC expense, to argue that the CBC had become too commercial and too American and, in the process, had lost its central place in the intellectual and artistic life of the country. One such group was the Committee on Television that included among others, Patrick Watson, nationalist scholar Abe Rotstein, filmmaker Allan King, and *Saturday Night* magazine editor Robert Fulford. It contended that the CBC's licence should not be renewed without a much stronger commitment to public purposes. Commercial advertising on television became a crucial issue. The CBC had already agreed to drop commercials on radio—a non-issue, essentially, because radio advertising by this time barely covered its own expenses—but, in the case of television, CBC President Laurent Picard flatly refused.

The "intellectual angels of English Canada," could rage as they liked, said Picard, but the CBC needed its commercial revenues and, without a commitment from the government to replace them, it would be "an incredible mismanagement of money," were the CRTC to remove them.[17] He also argued that "the freedom of the CBC" was at stake. When all is said and done, he insisted, the CBC reports to Parliament, not to the CRTC. The CRTC was only a paper tiger—a front behind which the government could hide.

And there the matter ended. The CRTC didn't really have the power to deny the CBC a broadcast licence. And the government, for its part, had no thought of paying for a non-commercial CBC, so it had no choice but to back Picard, which it implicitly did. The whole affair was laced with bitter ironies. One was that the CBC, which should speak for the public financing of television, was reduced to a heroic defence of its own commercialism on the ground that "the freedom of the CBC" was at stake. Another was that the CRTC, while attacking the CBC for failing to live up to the lofty aims of the Broadcasting Act, was simultaneously presiding over a major Americanization of the broadcasting system through the then-new technology of cable TV.

Cable created a dilemma for the CRTC. It had been set up to administer the 1968 Broadcasting Act, and the act was quite clear that the programming carried on Canadian systems should be predominantly Canadian. But cable allowed wider distribution of American television. This was inevitable along the border, where cable systems only provided a new means of delivering signals that were already available through the air. But the commission was soon faced with the question of whether it should allow these signals to be transmitted by microwave to cable systems in cities distant from the border. Initially, it said no. To actually build Canadian facilities to transmit American networks throughout Canada, said the commission, would represent the most serious threat to Canadian broadcasting since 1932. This brave stance provoked furious popular and political counter-pressure. It had always been the case that Toronto could watch American television, while Edmonton could not. But now technological means existed to remedy nature and history's unfairness in planting Edmonton farther from the border than Toronto. The CRTC never really had a chance. It had to capitulate, and it did so totally, without demanding any quid pro quo from the cable operators who would benefit.

The CRTC was in a box. Its instructions were to safeguard the Canadianness of Canadian broadcasting, but the only way to get anything more than token Canadian broadcasting was through public financing, and the CRTC had no power to create new public television. Nor had it much power over the well-established stations of the CTV network, whose prime-time schedules were packed with popular American offerings. So, it licenced new private television instead, starting in 1972 with the Global network. Global promised the CRTC the sky—Canadian programs from independent Canadian producers were the big selling point—and the commission's frustrations with its other charges, CBC and CTV, made it susceptible to this pitch.

History then repeated itself. Within three months, the infant network was facing bankruptcy. At that point, the CRTC could have withdrawn the licences that had not been fulfilled. Instead, it authorized its sale to new owners who rapidly turned it into another facsimile American network. Invested in its struggling offspring, the CRTC acted as the theory of "regulatory capture" predicts. In principle, the CRTC was responsible for the *performance* of broadcasters. In practice, it was responsible for their economic well-being.

During the heady days of the 1970 Canadian content hearings, it seemed as if the CRTC was well on the way to giving the broadcasting system the character that successive broadcasting acts have mandated: predominantly Canadian. Instead, the broadcasting system was further Americanized. Could the CRTC have turned this tide? The former commissioners I've talked to have generally said that this would have been impossible. Entrepreneurs wanted licences, audiences wanted American programs, and the government wanted an unrocked boat. The CRTC had neither the power nor the political standing to block their way. One former commissioner quoted the story of the French socialist, Jean Jaurès, who was one day standing with a companion in a Paris street when a large crowd ran by. Jaurès hurried after them, remarking to his companion, as he turned to go, "I must follow them—I'm their leader."

The consequences for the CBC were profound. It was forced to carry the entire burden of public purpose in an environment more and more inimical to that purpose. In the 1950s and into the 1960s, the CBC had tried to serve the full spectrum of Canadian taste: There was opera and Don Messer's Jubilee, hockey and Hamlet. But, after the 1960s, this range steadily shrank, as costs and competition increased. CBC Television didn't entirely lose its cultural ambition. The series, *Images of Canada*, which ran episodically between 1972 and 1976, took "the entire history of Canada" as its subject.[18] Nearly thirty years later, *Canada: A People's History* reimagined the country on an even bigger scale in seventeen dramatized two-hour episodes presented between October 2000 and November 2001. But the wild variety and spontaneity of the 1950s were mostly gone. Hugh Gauntlett, a department head at CBC Television between the 1960s and later 1980s, recalled that "rigorous lists of priorities and objectives" began to be drawn up in accordance with "modern management methods." These lists increasingly specified that "news and information is first … drama is second, and latterly, that everything else is, practically speaking, nowhere."[19]

Touchstone for the CBC

This was the situation Al Johnson tried to address when he took over from Laurent Picard in 1975. He was a career civil servant—like George Davidson before him, he came to the CBC from the Treasury Board—but he had a vision for the CBC. The title of the book he wrote about the CCF government he served in his native Saskatchewan, *Dream No Little Dreams*, could just as well describe his tenure at the CBC. Johnson's view was that the CBC must address what I have called the full spectrum of Canadian taste. *To accomplish this*, he thought, *it needed two things: more money and an additional television channel, so that one type of programming was not driving out another*. His argument, as he expressed it to me in an interview after he left the CBC, was as follows:

> Television is hamstrung because in the first place, we are partly commercial and we are partly public, tax supported, and you have mixed messages going to the producers of television. At one moment, you're telling them, produce a program that will garner large sums of money to enable us to produce Canadian programs, and in the next minute, you're telling them that we want you to do a thoughtful reflection on this subject or that subject. They're mixed messages. Secondly, the problem is that you cannot

> serve the regional and the national mandates of the Canadian Broadcasting Corporation on a single network. You haven't got the network time, the air time. And the third thing, and probably most important of all, is that you can't serve different audience tastes … With a single network, you're trying to serve everybody … When you have two services, you then are in a position to mold them so as to serve particular tastes in each of the two services. That doesn't mean that you have two different audiences. You have crossovers in audience, but you have the potential for a much richer service.[20]

At first, it appeared that this design could be realized. Thanks to the behind-the-scene work of his predecessor, Laurent Picard, Johnson assumed office with a promise in hand from the government of Pierre Trudeau that funding for the CBC would be guaranteed for five years and would increase by 5 percent in each of those years. This was a remarkable pledge, given the fact that the CBC had had to depend on uncertain annual appropriations ever since a licence fee model of funding for television was rejected back in 1952.[21] Johnson proposed to use this bounty to create a second television network and put forward a plan for how this was to be done in a document he and his staff prepared called "Touchstone for the CBC." CBC Radio already had a second network, Touchstone argued, and was using it effectively. Now it was time for television to do the same.

For a little while all stars seemed properly aligned. Then, suddenly, more urgent political priorities took over. In 1976, the *indépendantiste* Parti Québecois, under René Lévesque, was elected in Quebec. This set off a panic in the federal Liberal government of Pierre Trudeau, and the CBC's French radio and television service, *Radio-Canada*, was soon identified as one of the main sources of the government's problem. The attacks on *Radio-Canada* began in February 1977 with a speech in the Senate by former cabinet minister and Trudeau intimate Jean Marchand. "If ever this country is destroyed," Marchand told the upper chamber, "it will have been destroyed, in the main, by a federal institution that is financed by Canadians through their taxes." The next night in Toronto, another former minister, Mitchell Sharp, asked: "What is the CBC doing to help to break down the barriers surrounding the two solitudes and to promote harmony and understanding? My observation is bloody little. And I include both the English and French networks." There followed what *Maclean's* magazine called "an avalanche of bitter comments by cabinet ministers angry at the CBC and especially its French arm, Radio-Canada."[22] The capper came when Urban Affairs Minister André Ouellett told *The Toronto Star* that he had a list of "working separatists" at *Radio-Canada* and that he was prepared to share it with Al Johnson as a preliminary to having them fired. "Every night there's a bias," said Ouellett, "and every night it's in favor of the separatists."

Al Johnson fought back. He did not deny that there were people whose political views were separatist at *Radio-Canada*—"We have never engaged people on the basis of a political blood test," he said, "and we are not going to start now"—but he did deny that there was any bias in *Radio-Canada*'s reporting. After appearing before a committee of cabinet, and the Liberal caucus, to make his case, Johnson made a personal appeal to Pierre Trudeau to call off his dogs. At first, the prime minister appeared to comply, asking his cabinet colleagues, as *Maclean's* reported, "to back off their criticism." But, the very next day, in the House of Commons, he himself threw fuel on the fire. Liberal MPs, the prime minister said, "are very concerned about the possibility of CBC/Radio-Canada propagating separatism. I am not at all offended that they think so and that they say so." He even supported Ouellet's offer to name names. "I congratulate him for offering his services to the CBC," Trudeau said and then, warming to his task, he continued, "I do not think the opposition appears to be aware, and I am telling them with all the candour I can command, that almost everyone, including the high officials of the CBC, would be prepared to concede that the overwhelming majority of employees in the [French-language] CBC are of separatist leaning."

Not long after making this last statement, the prime minister threw the issue to the CRTC. But he didn't ask the commission to investigate the practical question of whether—and, if so, how—*Radio-Canada* was actually slanting its broadcasts. On that point, no evidence was ever adduced pro or con. Instead, he asked for an inquiry as to whether the CBC was fulfilling its legislative mandate, and particularly that part of it which required it, in the words of the 1968 Broadcasting Act, "to contribute to the development of national unity."[23] Harry Boyle, by then the chairman of the CRTC, took on the job himself and assembled a committee that reported later that summer. The Boyle committee tried to defuse and redefine the situation. The issue, it claimed, was not local outcroppings of partisanship but massive structural bias. The committee cited the work of York University political scientist Arthur Siegel who had recently done a large study comparing the content of English and French newscasts. Siegel investigated 1,785 news stories and found that only 259 had appeared in both French and English, of which half were international. In other words, the common ground between English and French newscasts was considerably less than 10 percent when it came to Canadian affairs. French and English, at least as far as information was concerned, were already living in separate countries. On this basis, the inquiry claimed that, yes, the CBC, and the media generally were biased—"biased to the point of subversion" was its dramatic claim—but that this bias mainly consisted of sins of omission rather than of active manipulation. "The mandate of unity," the committee wrote,

> has nothing to do with manipulating or distorting news, or inserting pro-federalist editorializing into the news. It is a very old principle that example is better than precept, and CBC television will do most for the unity of the country, not by editorially supporting federalism, but by regaining the presence in Canadian life that CBC Radio had a generation ago.

Harry Boyle had been part of the era in which radio had had a big presence in Canadian life, and he had watched in frustration from his seat on the CRTC as Canada's broadcasting system had inexorably slipped back into the condition the CBC had been supposed to remedy in the 1930s—American domination of Canada's airwaves. He was pleading, as Al Johnson was pleading, for a renewed commitment to the CBC. But his ingenious, and rhetorically adroit, attempt to reframe the instructions he had been given by the government fell on deaf ears. Al Johnson, instead of recognizing a friend, saw another enemy. He called the report "gratuitous and insulting." The rest of the media coverage that I saw kept the report within the frame of a government "witch-hunt"—Laurier Lapierre went even further and called it "terrorism"—without anyone ever seeming to notice what Boyle and his colleagues had tried to do. (I pointed this out at the time in an op-ed for *The Vancouver Sun*, but no one noticed that, in the way I had wanted, either.)[24]

The government boom fell on the CBC the following year. The promise of stable five-year funding with annual increases was rescinded and, instead, $70 million was cut from the CBC's budget. This then led to the final chapter of the tragedy, which was played out at the CRTC in 1981. I call it a tragedy, which to me it was, but, according to Marx's dictum that history happens twice, first as tragedy and then as farce, it might better be called farce, since the old tragedy of the CBC—fickle and conditional political support—was only being repeated.[25] The involvement of the CRTC was, likewise, a repetition. In 1974, the CBC had defied the CRTC; now it was the turn of the CRTC to frustrate the CBC. In both cases, the puppet master who held all the strings—the government—remained well-concealed behind the scenes. Al Johnson's dream of a second network could not come to pass without a broadcast licence, and that required CRTC authorization. But, by the time, the proposal went before the commission, the CBC plainly no longer had the political or—the same thing—financial support it needed to create such a network—a situation that made CRTC chair John Meisel acutely uncomfortable:

> My reaction was one of anguish and pain, essentially, because the idea of a CBC Two I thought was a wonderful idea. I couldn't have found [it] more appealing ... but there were some very, very serious problems with it, primarily, of course, the fact that there was no money for it. The money ... had to be taken from something else. And I found it very difficult to accept the idea that somebody like, say, myself, who is an avid radio listener, would find my radio service impoverished by money being diverted from it to enable cable subscribers somewhere, which really means the better-off people in the more urban areas of the country, to get ... an additional service ... And I thought that since the government was not prepared to put [in] additional funds ... it just wasn't appropriate that this be done. This was inimical to my mind to the kind of society in which we live and would undermine the quality of the CBC in places where it has a unique role, namely the areas where there is not much else.[26]

John Meisel and his colleagues felt, quite reasonably, that they were being asked to rob Peter to pay Paul—and, worse, to rob poor rural Peter on behalf of rich urban Paul. They refused. Al Johnson took a different view. He found the CRTC's rejection of CBC 2 "mind-boggling." What the commission should have done, he thought, was to have made common cause with the *CBC*. And then, he reasoned, "the government would have had to face up to the question as to whether it was going to veto the CRTC decision by refusing to give the CBC the money." This might have worked, and would certainly have been a brave try, but Meisel was trying to do his duty, as he saw it, in view of the CBC's actual circumstances, and not of Johnson's by then barely flickering hopes. The CRTC couldn't call the CBC to account in 1974, but it couldn't help it in 1981 either. The problem was structural—the CRTC, as I said earlier, was only a façade. The CBC was a creature of Parliament, as former President Laurent Picard had said. It had been given a legislative mandate and then denied the means, the political support and the broadcasting environment that would have allowed it to fulfill this mandate. If the government had wanted the CBC, in Harry Boyle's words, "to regain the presence in Canadian life that CBC Radio had a generation ago," then Al Johnson's Touchstone was as plausible and practical a plan for doing so as could have been devised. The commitment the government gave to Laurent Picard for stable, steadily increasing funding would have been enough to realize it. But the government's commitment was only a fair-weather pledge and couldn't withstand the first stiff political wind that tested it. Indeed, it's striking how casually and inattentively all this seems to have been done. In biographer John English's nearly 800 pages on Pierre Trudeau's career between 1968 and 1980, the affair doesn't rate even a mention. Public broadcasting just wasn't *that* important. The question that Harry Boyle vainly tried to raise in his report—what kind of CBC would be capable of generating real unity in the country?—was never addressed. Nor was the question of separatism at *Radio-Canada* ever intelligently addressed. There were certainly PQ supporters at *Radio-Canada*—André Ouellet's list, had he ever furnished it, would have had plenty of names on it—but the further question—didn't these people have a sense of their vocation as public broadcasters that went beyond their own partisan political commitments?—was never aired or even asked. How CBC/*Radio-Canada* discharges, or fails to discharge its public trust was a sleeping dog that the government preferred to leave asleep. The divided soul that Graham Spry imputed to R. B. Bennett also marked succeeding governments. In principle, Canada had an integrated broadcasting system under public control—the so-called single system. In practice, it had an American-dominated dual system whose eroding public component could be disowned whenever convenient.

The Liberal government of Pierre Trudeau had shown itself to be no better a friend to the CBC than earlier Conservative governments. Conservative critics had often focused on the CBC's supposed immorality. When CBC Television presented an avant-garde play called *Crawling Arnold*, by American writer Jules Feiffer, in February 1962, the Conservative member for Medicine Hat complained in the House of Commons that he and his constituents had found the show to be "depraved ... disgusting ... immoral ... garbage ... and a rank violation of the sanctity of the Canadian home and family."[27] His comments were of a piece with charges of communism, atheism, and sacrilege that had been issuing fairly regularly from the Conservative benches since the 1940s.[28] The attacks mounted from the Liberal side of the House by Marchand, Sharp, Ouellet, and Trudeau in the later 1970s were seemingly of a different kind. But whether the charge was treason or immorality, there was always a common element: the shallowness and inconstancy of support for the CBC.

The story of Touchstone's shipwreck on the sands of Canadian politics is, in this sense, a parable—a succinct summary of what has always and everywhere been the case. But, it was not, of course, the end of the tale. Johnson himself, though disappointed, still took considerable satisfaction in having steadily increased Canadian content on CBC Television. At the end of the 1960s, the prime-time schedule was evenly split between Canadian and American programs. By the time Johnson gave up his office to Pierre Juneau in 1982, it was three-quarters Canadian—a trend that has continued until today when the network is of almost entirely Canadian origin.[29] But, Touchstone had been playing for higher stakes—it had announced nothing less than "a battle for soul, for our cultural heritage, for our nationhood."[30] In these terms, it's hard not to think that, since Johnson's time, the CBC has been pushed farther and farther to the margins of Canadian life. This move has been slow, with many glorious exceptions and countercurrents, but it has been inexorable.

Fade to Black

The next epoch in this slide into marginality came in December of 1990 when severe budget cuts were imposed on the CBC by the Conservative government of Brian Mulroney. The CBC was forced to close eleven regional stations, eliminate the jobs of 1,100 of its 10,000 employees, and decimate local production. Several smaller cities lost the only local TV they had. Calgary, *Maclean's* reported, saw its "90-member staff … reduced to two reporters," and its local supper hour and late-night newscasts cancelled and replaced by broadcasts from rival Edmonton. Popular local shows like *Down to Earth* in Vancouver and the long-running *Land and Sea* in St. John's were dropped.[31] The carnage prompted Wayne Skene, then recently retired as director of television for British Columbia, to write "a requiem for the CBC" in a book called *Fade to Black* that he published in 1993. The CBC, he wrote, had been, "the nation's last, best hope for protecting the shaky elements of our culture and identity," but it had now "thrashed its way to the edge of irrelevance."[32]

One of the crueller ironies of this period was the role played by Patrick Watson. Watson had been CBC Television's *wunderkind*, its golden boy, and the face of its greatest popular success, *This Hour Has Seven Days*. In the wake of the *Seven Days* cancellation, he lobbied for the job of CBC president, and, in the years that followed, he created a lot of serious-minded television in series like *The Struggle for Democracy*, *The Watson Report*, and the historical vignettes he called *Heritage Minutes*. In 1989, he was made chairman of the CBC. Three years earlier a task force on broadcasting policy, headed by Gerald Caplan and Florian Sauvageau, had recommended dividing the executive authority formerly invested in the CBC president alone between the president and the chairman of the board. Watson was the first to occupy this new enhanced chairman's role, and he assumed it with what he thought were assurances that "enlarged resources" would go to programs. "I've had an unequivocal statement of support from the politicians involved," he said, "that they are completely behind the revitalization of the CBC."[33]

"The politicians," as it turned out, had led Watson up the garden path, as the cuts that followed showed. He ended up in an entirely untenable situation—caught in an uneasy relationship with CBC president, Gérard Veilleux, with whom he shared unclearly divided authority, and forced to answer for the decimation of the CBC's regional services. He attempted to salvage his hopes with a "summary paper" that he submitted to the board in 1994. In this memorandum, he argued that "the decline of political will to support CBC Television" was rooted in a "confusion about its role." At the heart of this confusion, he said, was the idea that CBC Television ought to out-compete private television networks on *their* terms by doing the same thing better. This was fatal, Watson argued, because "the only way CBC Television can survive is to become ruthlessly and unremittingly distinctive." This could be accomplished by drawing on "those members of the community of ideas, the policy elite, and the arts community who now feel neglected by public television in Canada." "A large reach and popular appeal" need not be sacrificed, he concluded, so long as programs are "outspoken, provocative and ground-breaking."[34]

Watson's proposal argued, in a sense, for a restoration of the formula that had made *Seven Days* such a success. CBC Television need not be torn apart by conflicting mandates, or divided into two services, in order to accommodate both popular and elite tastes. All that was necessary was for programmers to create an experience so compelling that all these oppositions would dissolve in it. "Make the programs good enough," Watson told his colleagues, and CBC Television could reclaim the attention of the country and recover its vocation to "always serve and never exploit." It was a captivating vision, but it implied something like the revolutionary spirit and receptive audience that had buoyed up *Seven Days*, and neither was much in evidence in 1994.

This is not to say that Watson's proposal left no trace. CBC Television in the 1990s did pursue distinction in the sense that it broadcast series like *Da Vinci's Inquest*, which had its own style and vision, and *Opening Night* in which performing arts usually excluded from network television, like dance, were showcased. This ended when Richard Stursberg took over as the head of English Television in 2004, hired by CBC President Robert Rabinovich. The appointment was unpopular at the CBC, where Stursberg was seen as an outsider, despite his having headed other cultural agencies like Telefilm and the Canadian Cable Television Association.[35] Rabinovich justified his controversial choice on the grounds that Stursberg had the right "skill set"—he was "creative," he had "experience in the industry," and he was "intelligent [and] a good communicator." National Arts Centre head, and CBC veteran, Peter Herrndorf added that he thought Stursberg's main asset was a "real sense of private sector flair [that] CBC [as] an organization needs."[36]

The Stursberg Era

Stursberg, as he later recounted in his memoir of his CBC years, immediately set about dismantling what he named—recalling Watson—the CBC's "distinctiveness strategy." According to Stursberg, this strategy reflected a perverse desire to "not do … whatever the private networks *were* doing."[37] This perversity arose, he said, from a prideful and contemptuous disregard for the public the CBC was supposed to be serving. Concerning "the attitude in the drama department," when he arrived at the CBC, he wrote: "The fact that Canadians yawned with indifference at their offerings confirmed that they must be on the right track. If the public was not watching, it must be because the shows were too demanding, too hard-hitting or too sophisticated. It could not possibly because they were badly made, poorly written or boring."[38]

Stursberg's caricature of the vain and self-defeating ways of his drama producers echoed the cocky, populist rhetoric that had animated *This Hour Has Seven Days* and the radio revolution. And this echo was quite intentional. After the 1960s, he claimed, radio had been "reinvented," while television had not. Radio had been transformed from "a snobby self-important elitist service to one that was much more informal, populist and engaged." Now, at last, it was television's turn to undergo a comparable "democratic revolution:"

> The Radio Revolution was a dramatic change in tone and manner. It recognized, as television had great difficulty doing, that society had changed. The old distinction between the high arts and popular culture had eroded to the point of meaninglessness. There was no cultural elite whose job it was to educate or lift up the sad unwashed. There were only Canadians, all equally interesting, funny and important. The Radio Revolution was fundamentally a democratic revolution. It embraced everyone.[39]

Stursberg was ostensibly repudiating Watson's "relentless and unremitting distinctiveness," but the two actually had quite a lot in common, at least in terms of their rhetoric. Both claimed that sheer, compelling quality could dissolve old contradictions between genres and audience segments. Both wanted to provide CBC audiences with an integrated experience—Watson spoke of making a world so consistent that the viewer would find it impossible to withdraw; Stursberg, with his "private sector flair," offered "a culturally coherent brand promise."[40] Both were intoxicated by the democratic potential of television.

Soon after Stursberg arrived at the CBC, all Toronto staff were summoned to a large television studio, linked to CBC locations across the country, and the new order was unveiled. CBC Television would weed out all underperforming television programs. Series like the two I mentioned earlier—*Opening Night*, and *Da Vinci's City Hall*, a sequel to *Da Vinci's Inquest*—both of which were getting rated audiences in the 200,000—300,000 range, were immediately cancelled. A target of 1,000,000 was set as a threshold for success. A new slate of programs was announced. The most heralded of them were given "unprecedented promotion budgets"—*Little Mosque on the Prairie,* for example, was launched with "camels in Dundas Square [and] falafels … handed out at Union Station." The first broadcast of *Little Mosque* in January of 2007 had a rated audience of 2.1 million viewers, and the show's audiences continued to exceed Stursberg's threshold for the show's first two seasons, declining thereafter. Several other hits were also registered. There was *Battle of the Blades*, a figure skating contest in which elite female skaters were partnered by professional hockey players, and the still continuing *Dragon's Den* in which aspiring entrepreneurs run a gauntlet of celebrity investors.

Television, according to Stursberg, is inherently an entertainment medium, and he set out to "respect the medium"—"[when] the day is done," he wrote, "Canadians [are] ready to be entertained."[41] On these terms, he succeeded—a number of the programs brought to air during his six-year tenure were considerably more popular than what was on the air when he started. This, however, did not prevent his being abruptly fired in 2010. Within the Toronto Broadcasting Centre, where I worked at the time, it was widely believed that he had been "walked out," though this was denied by the new CBC President Hubert Lacroix. Lacroix claimed, in a carefully coded all-staff memo, that the sudden firing "was, rather, the culmination of a very long reflection on the future of the Corporation, the culture it needs to adopt in order to change and adapt in an evolving media environment and our ability to agree to a long-term plan based on a shared vision,"[42] He then went on to affirm "that there is nothing (and I mean nothing) in our current programming strategies that I don't stand by: so, those out there who think this is in any way a repudiation of where we stand today will be disappointed big time." The two statements—I want to change "the culture;" I totally back "our current programming strategies"—seem diametrically opposed, but are not actually so. Stursberg's acerbic, outspoken style had rubbed people the wrong way, but his revolution had taken. "Elitism" at CBC Television had been purged at last, and the television service had successfully "made the transition to a content company." What was easier than to sacrifice the innovator, ostensibly to heal "the culture," while retaining the innovation?

Television Today

I have watched many of the dramatic series that CBC Television has broadcast in recent years. The first thing I have learned is how extraordinarily difficult it is for the CBC to reach a wide audience. Even the successes—*Little Mosque*, say, or, more recently, *Schitt's Creek*—seem to command only a fickle and fitful attention. Many series pass almost unnoticed—at least in my milieu. Friends with whom I share an interest unconnected to the CBC seem generally unaware and uninterested in what Canada's public broadcaster currently has on offer. I have heard *The Wire*, *The Sopranos*, and *Breaking Bad* enthusiastically discussed at the bar after hockey but never a CBC Television series. Second, I have been forced to the conclusion that many of these series have a generic or modular quality. They are, perhaps, what you would expect from a "content company"—works created to fit a format or confirm a prejudice rather than expressions of some urgent inspiration. This is not at all to say that they are badly done. Often enough they are well-cast, well-acted, handsomely shot, nicely paced, and smartly written. It's just that they seem to have been devised according to a formula for dramatic writing. They lack any heart. Perhaps this arises from what Stursberg called "respect [for] the medium itself." The shows are written to accommodate short attention spans and frequent, prolonged commercial interruptions, and they deploy already prevalent, and, for this reason, quickly and easily grasped stereotypes in order to make an immediate impact. Characters are endowed with an attractive appearance of individuality by skilled actors and skilled writers, but this individuality has no depth. There's no one there that you could get to know or whose acquaintance might, in some way, nourish or instruct or fortify you as a viewer. There's no *there* there.

Here's an example. In January and February of 2020, the CBC presented an eight-part dramatic series called *Fortunate Son*. A period piece set in 1968, the story concerns a family in British Columbia who help American draft resisters to escape to Canada. The parents are both American exiles who have been involved in violent political action before escaping themselves. The plot is involved, and I will not attempt to summarize it, since it is beside my point. I was attracted to this series by its subject. I had been connected to American draft resisters in Toronto in those years, and to the organization and the journal—AMEX, for American exiles in Canada—that they created. I have also long been interested in the way the legends of the American 1960s have tended to eclipse and even erase the memory of those years in Canada. I watched attentively. At first, it seemed as if the setting in which I was interested was just being used as a colourful background for a thriller, full of half-explained skulduggery and up-in-the-air episode endings, but gradually I became attached to the characters. The actors and the photography were excellent and the writing was efficient and evocative—each scene drawn in a few telling strokes—probably allowing my imagination to supply more than was really there. But, after this world had been created and brought to life, it was then brought to an end so quickly and with so little regard for its characters, that I felt cheated, betrayed, and slightly brutalized.

The problem wasn't that an implausibly happy, if still suitably ambiguous, ending was provided. This might be said of many comedies—that they're just tragedies rescued by their authors at the last minute. It was that the end came so abruptly, and with so little regard for the meaning of the events previously enacted, as to make it seem that the characters had only been paint-by-numbers all along. The scene had been no more than an exotic locale, pasted together from the clichés of the American '60s. No attempt had been made to remember the Canadian '60s and their quite different tenor at all. The series had been a triumph of format over art; of cliché over living history; and of entertainment over genuine storytelling. The characters had never exceeded their stereotypes, even if my carefully manipulated desire had sometimes made it seem as if they had.

How impersonal the CBC's entertainment machine has become is illustrated by the series' presentation. *Fortunate Son* received no introduction—Hockey Night in Canada still has a host, the news still has an anchor, but dramatic series just appear and disappear, unannounced, unexplained, and uncontextualized—phantasms in a void, neon lights blinking on a commercial strip. This was egregious in the case of the closing credits, which scrolled by in a hurried, illegible blur of tiny print at the bottom of the screen while a hyperactive promo played above them on the right and a second-by-second countdown to the nightly newscast, *The National*, appeared on the left. (Such haste and disregard is now standard on CBC Television.) The effect on me is of a loud announcement that none of this really matters, that none of the people who created this program are anything more than cogs in the machine.

As I said, a certain heartlessness is involved. The programs are the product of a carefully contrived method, rather than of some artistic necessity or personal vocation. (Or if there is a personal vocation, since in the case of *Fortunate Son* the central character was based on the mother of one of the executive producers, this vocation is overwhelmed by the generic elements of the undertaking.) The production is pulling strings rather than addressing the viewer. Attention is being adroitly gained, held, and extended but it is not being nourished. The characters are finally disposable. And just as the producers are practised in making their fast food tasty, so does the audience acquire skill in being used in this way. It becomes a mark of sophistication to recognize, enjoy, and, at the same time, distance oneself from the experience of being manipulated.

I mention *Fortunate Son* only as an instance of something much more general. An earlier dramatic series, *Pure*, which ran on CBC Television in 2017 and 2018 had many of the same attributes—intriguing, appealing, and ultimately empty. In its case, a Mennonite setting was used with the same disregard for its significance as *Fortunate Son* showed for its time and place. *Anne with an E*, a series that ran for three seasons between 2017 and 2019, showed the same disregard in a somewhat different way. Equally accomplished as the other programs I have mentioned and blessed with three superb actors in its principal roles, it betrayed its place, and its literary original, by turning the past into the present.[43] Using the characters and setting created by L. M. Montgomery, the creators of *Anne with an E* turned their protagonist into a fully woke modern gal and set her loose on turn-of-the-(last)-century Prince Edward Island. In this case, ideology rather than format banished reality. *Anne* freed its viewers from the past, by replacing the past's obligations and constraints with a fantasy of what should have been. "Not even the gods can change the past," says Aristotle, but now the CBC can.[44]

Richard Stursberg made it a point of honour for CBC Television to begin producing programs that people wanted to watch. He made fun of the effete drama producers who thought—so he said—that failure was success, and of the self-indulgent elitists who saw "large audiences" as a sign that the work "must be vulgar and stupid."[45] Popular success, as he portrayed it, is a kind of ordeal to be undergone for the common good, a gauntlet that he dared his sheltered colleagues to run. It was a bracing vision in some ways, and it echoed many past CBC executives, from Davidson Dunton to Al Johnson who had resisted becoming what Johnson called "a boutique broadcaster."[46] But, in the end, it was just another twist of the sterile populist/elitist dichotomy. And it has worked out badly, unmooring the CBC from its heritage and its proper dignity, and providing only fleeting and insubstantial success in return. Battered by political neglect and unrelenting commercial competition, it has abandoned its vocation for a will-o'-the-wisp.

The National Question

In 1965, philosopher George Grant, in his celebrated *Lament for a Nation*, proclaimed the end of Canada as a distinct and distinctive political formation. The country from which his ancestors had looked askance at the "licentious republic" next door had become, he said, just another standard issue, American-style technological society.[47] His widely read book had a paradoxical effect, reviving hope for what the book claimed was now beyond hope, and stimulating a new era of nationalism among artists and scholars, political organizers, and CBC producers. This revived nationalism had many expressions. The performing arts blossomed; Canadian theatres began to present original Canadian plays; Canadian literature acquired a more assured voice and a new international stature; the Watkins Report and books like Kari Leavitt's *Silent Surrender* galvanized opposition to American domination of the Canadian economy; the Waffle movement stirred up the NDP.[48] The CBC was very much a part of this movement of self-discovery. In television, I think of director Vincent Tovell's 1975 program *Journey without Arrival*, which explored Northrop Frye's vision of Canada, and of *A Gift to Last*, a dramatic series that ran in 1978 and 1979. Starring Gordon Pinsent, who also contributed to the creation and writing of the series, and set in rural Ontario at the time of the second Boer War (1899–1902), *A Gift to Last* embodied many of the virtues that I find lacking in today's dramatic offerings. A contemporary sensibility, I imagine, might find it a little bit clunky—naïve perhaps—and lacking that high gloss, hyperactive pace, and cynical sophistication that viewers are now thought to expect. But what marks *A Gift to Last* out for me, and makes it a foil for the entertainments I was criticizing earlier, is the sense of place and the obvious affection with which I found it to be imbued. It brought before me a country I wanted to know about, presented characters that *did* exceed their stereotypes, and gave me a sense of participation in something that inspired love and hope.

"It all turns on affection … Affection. Don't you see?" says Margaret Schlegel to her hard-headed husband Henry at the culmination of E. M. Forster's novel *Howard's End*. George Grant often spoke, in the same vein, of "love of one's own"—the love of what has been handed over and entrusted to us, whatever its character, just because it is ours.[49] The CBC, during this flowering of nationalism that I have been talking of, sometimes embodied this spirit: There was a sense of coming to know and love this place. But the struggle was always uphill. One memory stands out particularly for me. In 1976, as I've said, I was one of the hosts of *Good Morning Radio*, the CBC's morning show in British Columbia. In this connection, I was given an advance screening of a film called *Dreamspeaker* by Quebec director Claude Jutra that was to be shown on CBC Television the following Sunday evening. Jutra was then one of the young lions of Canadian film-making, having already enjoyed great success with *Mon Oncle Antoine* (1971), which he cowrote and directed, as well as receiving critical (but not commercial) acclaim for *Kamouraska*

(1973), a panoramic imagining of Anne Hebert's novel of that name set in 1830s Quebec. I was thrilled by *Dreamspeaker*, and by the performance of Tseshaht actor George Clutesi as the shaman, or dreamspeaker, of the title, and I saw the film as another epoch in the discovery of Canada that I've been talking about. I concluded Friday's broadcast of *Good Morning Radio* with as urgent and as fervent a plea as I have ever made for people to watch it. The following Monday morning, back in the office, I quizzed my colleagues about how they had liked it. Not one had seen it. It turned out that a blockbuster American network series called *Roots* had been on at the same time. (*Roots* was based on a novel by Alex Haley, which told the story of an enslaved West African and his descendants.) This preference for the American program may not seem particularly remarkable today, when Canada's status as an American subculture is widely accepted, but the memory still has a bitter taste for me—not least because *Roots* was, I'm sure, just as significant *for Americans*, as I had hoped *Dreamspeaker* would be for Canadians.

Other memories fall together with this one. In 1983, director Robin Phillips made a film of Timothy Findley's novel *The Wars* with a who's who of Canadian acting talent in the cast. *The Wars*, though utterly different from *Dreamspeaker*, had for me the same ur-Canadian quality that I had discerned in Jutra's film—it was another answer to Northrop Frye's question, where is here?[50] The film was sombre, like the novel, but powerful. I watched it in a nearly empty cinema on St. Clair Avenue where it closed after only a week.

In 2021, we have an almost entirely Canadian CBC Television telling ostensibly Canadian stories. But they are mostly generic stories in which Canadian people and places are just colourful local outcroppings of Global Entertainment. The hopes of my youth—that we would finally live in a distinct and unique *here*, aware of its traditions, knowledgeable about its history, and protective of its institutions—are gone. The CBC was part of this hope and has been part of its failure. There have, of course, been magnificent exceptions, like the thirty-two hours and seventeen episodes of *Canada: A People's History* in 2000 and 2001. Another, worthy of note, was *The Valour and the Horror*—three films by Brian and Terence McKenna that presented the shadow side of the Second World War's reputation as the good war. This series was controversial, and a number of veterans, including members of my family, mobilized against it.[51] I applauded the CBC's decision to commission and broadcast these programs. Canadian airmen participated in raids whose deliberate purpose was to create firestorms in German cities—a patent act of terror and genocide. Asserting this reality against the reverence that had long kept it out of sight and out of mind seemed to me an exemplary exercise of public purpose. The critics claimed, with some justice, that the filmmakers ought to have presented their findings more tentatively as their *account* rather than as an authoritative retelling, but I doubt if any way of telling the story would have averted the anger that followed. What was terrible about the affair for me was that the CBC reacted by putting *The Valour and the Horror* on the shelf. The series was never rebroadcast, and the issues it raised got no further discussion on CBC Television outside of brief news snippets. Instead of pursuing the issue, by inviting further investigation and discussion, the CBC timidly ducked and covered. What should have been the beginning was the end. The CBC, for a moment, had stepped into its proper role as Canada's public forum, and then immediately stepped out again—an image of the disappointed promise I've been describing.

Scholar Marc Raboy called his history of Canada's broadcasting policy "missed opportunities"—an apt and telling phrase that Raboy quotes from Graham Spry.[52] I have drawn attention here to a few of the missed opportunities that have stood out for me. The result has been that CBC Television has ended up in a cul-de-sac, a street with no exit. For nearly seventy years, between 1936 and 2004 the CBC in some way sustained the tension which its contradictory purposes make inevitable. Then Richard Stursberg gave the order to relax. There *was* no "cultural elite," nor any meaningful distinction between "the high arts and popular culture." There were "only Canadians." As the tension relaxed, the air went out of the CBC. It's now essentially a private network with a subsidy, just as its critics say, and the very meaning of *public* broadcasting has grown obscure.

Epilogue to Part One: The CBC and the Pandemic

> Complete liberty of contradicting and disproving our opinion is the very condition which justifies us in assuming its truth for purposes of action; and on no other terms can a being with human faculties have any rational assurance of being right.
>
> —*John Stuart Mill,* On Liberty[1]

> The right to search for truth implies also a duty; one must not conceal any part of what one has recognized to be true.
>
> —*Albert Einstein*[2]

I ended Chapter Five with some reflections on the current homogeneity of opinion at the CBC. This was well, and decisively, illustrated for me by the CBC's reaction to the COVID-19 pandemic, and so I'll conclude this section of the book with a brief history of this event and the way in which the CBC responded to it.

In March 2020, the World Health Organization (WHO) declared that a pandemic was in progress. The announcement seemed to electrify and align entire populations at a stroke—just when critical thought was most needed, it was instead suspended. The CBC participated enthusiastically in this orgy of unanimity and has continued to do so ever since. This might be expected from private broadcasters whose business is to gratify their listeners. But, public broadcasting, ideally, has a different vocation. It exists to discern the public interest, as successive broadcasting acts and CBC mission statements have said. What the public interest demanded at the beginning of the pandemic was careful deliberation. What the CBC delivered was thoughtless cheerleading.

In March 2020 little was known about the illness that the WHO said had become a "pandemic"—a term which implies, to the common ear, a deadly plague, even though it has no agreed definition other than widespread illness. In the absence of any actual knowledge of the infection mortality rate—that is, how many of those infected would die—terrifying speculative models circulated. The most pervasive was hurriedly produced by the COVID-19 Response Team at Imperial College London. This model tried to predict what might happen in three possible cases: no intervention, moderate intervention, and aggressive intervention. In the first case, the forecast was a disaster: 2.2 million deaths in the United States, more than half a million in the United Kingdom, and so on. The second case was also pretty bad, but the predicted outcome with aggressive action was much better. This model had less foundation than the average weather forecast, since the disease was new, and little was definitively known about either its virulence or its communicability. Nevertheless, the model's predictions quickly carried the day. "I don't think any other scientific endeavour has made such an impression on the world as that rather debatable paper," stated Johan Giesecke, a former chief medical officer in Sweden.[3] Some eminent epidemiologists, like medical statistician John Ioannidis at Stanford, warned of what he called "a fiasco in the making" if drastic action were taken in the absence of either information or deliberation.[4] In Canada, Dr. Joel Kettner, a former chief medical officer of health in Manitoba, issued a similar warning. He phoned CBC Radio's *Cross-Country Checkup* on March 16, 2020, to point out that "social distancing" was a largely unproven technique. "We actually do not have that much good evidence," he said. "It … might work, [but]we really don't know to what degree and the evidence is pretty weak." These voices were not heeded, or even much listened to. In the case of Dr. Kettner, *Cross-Country Checkup* host Duncan McCue listened with strained courtesy, then let the matter drop.[5]

It was quickly understood that a policy of universal quarantine, or lockdown, must be pursued. Those who resisted, as Britain did for a short while, were mocked and bullied into line. In the end, only Sweden managed to hold out, and, even then, only partially, against this policy. This was surprising in several respects. First, there's the word *lockdown* itself. It was used first in prisons—to describe inmates confined to their cells after a disturbance—and then occasionally applied in large schools where students were temporarily shut in after some violent incident had occurred. How this harsh word suddenly began to fall so gently on the public ear, becoming the name of a benign expression of mutual regard for which all right-thinking citizens clamoured, is a study in itself. But even more startling was the complete reversal of what had formerly been axiomatic in the field of public health—that the public should not be unduly alarmed, that quarantine was a tool to isolate the sick and shelter the vulnerable, and that *public* health should be concerned with the general good and not just the management of a single disease vector. Now COVID-19 eclipsed all other concerns; quarantine was for the entire healthy population; politicians and public health experts vied with each other in frightening the public; and, most remarkable of all, this revolution passed unremarked, as if it were simply common sense and not a dramatic departure from existing norms. The prime minister, seemingly without a thought for the terrors of cholera or smallpox, declared that we were in "the greatest health care crisis in our history."[6] Public health experts predicted that two or three out of every hundred infected might die. The media spoke of nothing else. This response was typified for me by a banner headline in *The National Post* that seemed to function more as an instruction than a description. It said simply: PANIC.

A war mentality settled in, and, with it, a disposition to divide society into friends and enemies. *The Globe and Mail* made it explicit in a September 2020 editorial headlined "Forget politics: It's time to fight COVID," which declared bluntly that "Canada is at war," but, by then, the metaphor and its cognates were already everywhere.[7] This meant, among other things, that information was selected according to its conformity with our war aims. This was sometimes called "staying on message" in keeping with the idea that in times of war the task of the media is propaganda. The story, and the mood, established at the outset must be sustained, whatever new information became available. So, for example, when John Ioannidis was finally able to estimate the infection mortality rate, he arrived at a median rate of 0.27 percent, a figure that fell to 0.05 percent when deaths among those over seventy were excluded.[8] This was peer-reviewed research, published in the *Bulletin* of the WHO, which showed that the likelihood of death was less by many orders of magnitude than the original frightening predictions, and yet the news was barely reported. The same pattern held with the fear that the hospitals and the health care system generally would be overwhelmed. Every time COVID-19 cases "surged," the hospitals "braced" for the onslaught, and then, when the hospitals were not overtaxed, it was never mentioned. It was odd, in the first place, that during the "greatest health care crisis in our history," the citizens were charged with "protecting" the very system that ostensibly protected them, but the constant reiteration, without evidence, of the image of the about-to-be-beleaguered hospital made it worse. It also effectively hid the question of whether the hospital was the best place for many of the people who *were* there. How many suffered in frightening and debilitating isolation in hospitals who could have been better cared for at home, with only minimal medical support?

The CBC took on its propaganda role with the same alacrity as the newspapers I have quoted. I will illustrate with an example that continues to shock and embarrass me to this day. On March 22, 2020, CBC News Network interviewed Dr. Richard Schabas. Schabas had been Ontario's chief medical officer of health between 1987 and 1997 and had appeared, or been heard, on the CBC, by his estimate, "literally hundreds" of times before. For more than thirty years, he had been a regular and trusted source on public health matters, and he had fully justified the CBC's confidence in him. During the SARS outbreak in 2003, for example, Schabas was the chief of staff at one of the affected hospitals, York Central. At a time when there was widespread panic and some models were predicting 120 million deaths worldwide, he determined that the disease was not sufficiently infectious to spread in community settings and predicted that it would die out as soon as proper infection control measures were adopted in hospitals. He was proved right.

In his interview on March 22, 2020, he questioned some of the measures, like lockdown, that were being taken against COVID-19. The interview disappeared from the CBC's website the day it was posted. CBC News Managing Editor Tracey Seeley wrote to her colleagues: "NN [News Network] unfortunately ran an interview with Dr. Schabas this morning, and a clip was included in our web story. We … had Uncoder unpublish it completely. [Such] sources are considered 'the climate change denier' equivalent of corona prevention." Dr. Schabas, Seeley said, was an "outlier" and "should be treated as such." Schabas never heard from the CBC again—he had been cancelled. He later made a complaint to the CBC Ombudsman, in which he produced the emails I have quoted, and argued—I think quite justifiably—that, by his cancellation, the public were being denied "a breadth of knowledge and experience that is virtually unique in Canada." His complaint never received a substantive answer.

The careful and considered opinion of a seasoned professional who had proved prescient on public health matters in the past was suddenly equivalent to "climate change denial." This set the tone for coverage in which scientific disagreements were rigorously excluded from public discussion. People were simply told that their leaders were "following science," regardless of the fact that the available science was both scant and contested. The virus was new, so was the policy being used to combat it, and yet "the science" was held up as if it offered transparent, obvious, and unequivocal guidance. Often the term meant nothing more than the opinion of some credentialed person.

The question of the utility of masks provides a further example. No randomized controlled trial has ever shown that masks—even good masks properly fitted—reduce transmission of respiratory viruses. At the beginning of the pandemic, in April 2020, retired Canadian physicist Dennis G. Rancourt surveyed the existing scientific evidence in a paper for the Ontario Civil Liberties Association and concluded unequivocally that "masks don't work."

"There have been extensive randomized controlled trial (RCT) studies, and meta-analysis reviews of RCT studies," he wrote in his abstract of this article, "which all show that masks and respirators do not work to prevent respiratory influenza-like illnesses, or respiratory illnesses believed to be transmitted by droplets and aerosol particles."[9] A more recent meta-analysis of seventy-eight randomized trials, reported by the Cochrane Library in January 2023 confirmed this finding. "The pooled results of RCTs," wrote the thirteen authors of this systematic research review, "did not show a clear reduction in respiratory viral infection with the use of medical/surgical masks." "Wearing masks in the community," they conclude, "probably makes little or no difference to the outcome of laboratory-confirmed influenza/SARS-CoV-2 compared to not wearing masks.[10] Various observational studies were reported during the pandemic that did accord masks some benefit, but it has long been an axiom of the movement for evidence-based medicine that randomized trials, by eliminating selection bias, provide the highest standard of scientific evidence. So, although further study might change this conclusion, and there is still "uncertainty about the effects of face masks," according to the Cochrane Library study, it is currently the case that masks in community settings have *no proven efficacy*. There is, moreover, the still largely unstudied question of their physical and psychological side effects. These range from impeded breathing and inhalation of microplastics to the climate of threat and mutual mistrust that masks potentially create.

Now consider what happened during the pandemic. When it began, both the WHO and Theresa Tam Canada's chief public health officer took the view that Rancourt stated so bluntly in his study for the Ontario Civil Liberties Association: Masks don't work. Both then changed their view without any change in the evidence. The only explanation for this, that I can think of, is that there was a desperate need for a way of ritualizing the fear that had been generated, and masking provided the needed ritual. This is understandable. But what is remarkable is how easily this rain dance then became "the science." Rancourt's study was easily available, but it was never, to my knowledge reported—on the CBC or anywhere else. Instead, the "anti-masker" became the very epitome of anti-scientific bigotry.

Throughout the pandemic, scientific dissent about all aspects of COVID-19 policy was systematically excluded from the CBC. What began with the banning of Richard Schabas as an "outlier" continued thereafter. This was most egregious in a case that occurred in the summer of 2020. It was then that Drs. Schabas and Kettner—the former Manitoba chief officer of public health—joined with a number of other former public health officials to question the policy of universal quarantine that all Canadian governments were then following. This distinguished group included, along with the two former provincial chiefs, two former chief public health officers for Canada, three former deputy ministers of health, three present or former deans of medicine at Canadian universities, and various other academic luminaries—a virtual who's who of respected elders in the field of public health in Canada. In their open letter to the prime minister and Canada's political leaders, they pleaded for "a balanced response" to the pandemic, arguing that the "current approach" posed serious threats to both "population health" and "equity."[11] To the best of my knowledge, this letter was never reported on the CBC or in the press and never answered by any of the political leaders to whom it was addressed. It was months after it was written that I even heard of it—through a friend. Again, I can only underline how extraordinary, and how ominous, I find this to be. Cautions against lockdown and other related elements of COVID-19 policy, from what had recently been the public health establishment, ought to have raised alarm in both political and media circles, and certainly at the public broadcaster. What followed instead was silence.

The same story was played out again a few months later when three eminent epidemiologists made what came to be called the Great Barrington Declaration, after the town in Western Massachusetts where it was issued.[12] The authors were Martin Kulldorf, a professor of medicine at Harvard, Sunetra Gupta, a professor of theoretical epidemiology at Oxford, and Jay Bhattacharya, a professor of medicine at Stanford. Their statement deplored "the devastating effects on … public health" of the total mobilization then underway and advocated instead what they called "focused protection"—a policy of protecting those at risk from COVID-19 while allowing everyone else to go about their business. In this way, they reasoned, immunity could gradually build up in the healthy population, without endangering those who were particularly vulnerable to the disease. Their declaration was ignored by the mainstream media in Canada and vilified in the war rooms of pandemic policy. The head of the American National Institutes of Health (NIH) Francis Collins, concerned that the declaration was getting "a lot of attention," via new media, wrote to Anthony Fauci and several other NIH colleagues, calling for "a quick and devastating take-down of its premises." The three authors, Collins said in his email, were "fringe epidemiologists."[13] The word "take-down" from this ardent defender of "science" is telling. The CBC—again so far as I can find out—never mentioned the affair.

The crucial point here is that the pandemic provoked extensive and deep-rooted scientific disagreement, but news of this disagreement never reached the public. The CBC established a strict and effective censorship, as the case of Richard Schabas's cancellation shows. Those who tried to resist this policy were silenced. One who has told her story is Marianne Klowak, a thirty-two-year veteran of the CBC Radio newsroom in Winnipeg.[14] She has said that she was prevented throughout the pandemic from covering any story relating to protest against COVID-19 policies, scientific dissent, or vaccination side effects. When she argued that the CBC ought to try and find out something about the large peaceful protests that were assembling outside the Manitoba legislature, she was told that these were "anti-vaxxers" and therefore not deserving of a voice. On another occasion, she was, in her words, "gaslighted" in front of her entire newsroom—that is, made to feel crazy for persisting in her desire to cover the other side of an issue that her colleagues insisted had no other side. Her attempts to cover cases of vaccine injury were blocked. Eventually, she left the CBC.

One of the ways in which the policy of censorship was justified was by characterizing all dissent from the principal elements of COVID-19 policy—lockdown, compulsory masking, vaccine mandates, etc.—as *misinformation*. In the summer of 2019, the BBC initiated what it called The Trusted New Initiative, an undertaking that soon enlisted an elite group of media organizations including digital giants like Facebook, Google, and Microsoft. The stated purpose was to "protect audiences and users from disinformation." The CBC quickly signed on. A subsequent "summit" meeting of the partners in this initiative declared that "vaccine disinformation" would be a particular focus.[15] The statement released by the summit made a distinction between "legitimate concerns about vaccination" and "harmful disinformation" and said the partners would respect the former. But legitimacy is in the eye of the beholder, and facts are not always easily separated from the theories that make them facts in the first place. In the event disinformation, or more commonly misinformation, often came to mean disagreeable opinions and not demonstrably false statements.

YouTube was particularly aggressive in banning videos whose point of view it found uncongenial, claiming its "terms of service" had been "violated." In one case, it took down a video of a round table discussion convened by the Governor of Florida, Ron DeSantis, between himself and a group of scientists, which included the three Great Barrington authors. The governor's interlocutors argued that "focused protection" would have been a more intelligent, more effective, and much less destructive response to the pandemic than the policy of mass quarantine—lockdown—that was actually followed. An email statement from YouTube, which is part of the Trusted News Initiative, explained that the offending discussion had been removed because "it included content that contradicts the consensus of local and global health authorities regarding the efficacy of masks to prevent the spread of COVID-19."[16] Further evidence of government censorship of American media was provided when Elon Musk released the so-called Twitter Files, following his purchase of the social media company in 2022. These showed that Twitter executives had held regular discussions with officials of the FBI, the CIA, the Pentagon, the State Department and other government agencies, which resulted in censorship and suppression of Twitter accounts.[17] Great Barrington author Jay Bhattacharya found that he had been put on a so-called trends blacklist, which prevented people from outside his network of followers from ever becoming aware of his tweets.[18] In Canada, veteran reporter Rodney Palmer has shown that virtually all the "independent" scientific experts on whom the CBC relied during the pandemic were receiving funding from *Science Up First*, a Canadian government initiative designed to "stop the spread of misinformation."[19] It was on these grounds—stopping the spread of misinformation—that Marianne Klowak, at CBC Winnipeg, was prevented from talking to anyone outside the bounds of the official COVID-19 consensus. The CBC, in this case at least, functioned in exactly the way the opponents of public broadcasting claim that it does—as a mouthpiece of government.

By this suppression, or mischaracterization, of all dissent, the public was brought to believe that an indisputable oracle called *Science* was issuing unequivocal advice that need only be "followed." Had the many sciences bearing on infectious viral diseases really been unanimous, and had COVID-19 really constituted "the worst health care crisis in our history," as the prime minister averred, this might have been understandable. But the scientists, as I have shown were far from unanimous, and the real severity of the health crisis undergone by Canada was hard to determine without free inquiry and free discussion. What does seem certain to me is that great harm was done both to science and public policy by the propagation of an image of science as a unified and consistent agency offering unquestionable advice. This image hurt science by mystifying its character—in place of fallible, contestable, plural *sciences*, people were offered a reactionary idealization. It hurt public policy by disguising the unavoidably moral and political character of the decisions that were being made. "Science" had decided to confine healthy people in their homes—who dared disagree? The resulting harm has yet to be calculated. When it is, it will have to include, at the least, lost livelihoods, failed businesses, astronomical debt, epidemic fear, untreated illness, social demoralization, and all the ills fostered by isolation. Were these foreseeable harms a reasonable price to pay in order to "flatten the curve" and "protect our health care system"? The question was never raised in the organs of polite opinion—one was following science. Even populist politicians meekly submitted—"I gotta do what the docs tell me," said folksy Ontario Premier Doug Ford.

It has often been noticed that an atmosphere of crisis allows the introduction of changes that would normally be resisted. In the case of the pandemic, people who would never previously have let a morsel of genetically modified food pass their lips uncritically accepted a vaccine employing an untried genetic delivery system. Wary opponents of Big Pharma became credulous and indulgent friends. "A new normal" was said to exist. What had seemed bedrock principles disappeared overnight—most notably the principle of informed consent that was enunciated in the 1947 Nuremberg Code as a response to the abuses committed by Nazi doctors. Informed consent, throughout my life, has been presented as the very foundation of the legitimacy of contemporary bio-medicine. Vaccine mandates abolished it overnight without debate or discussion. Amazon thrived, while Main Street suffered. The virtualization of work took a great leap forward. Dissident opinion was mocked and marginalized, creating an ominous social polarization. Bonds between individuals and the state tightened, while bonds between people weakened. Safety, bio-security, and control of risk established themselves as the ultimate political necessity and raison d'état—a necessity stronger even than the now revocable right to console the dying or gather for religious celebration.

One doesn't have to be a conspiracy theorist to discern in these changes the outlines of a new social order. This order could not have arrived with such seeming suddenness, had it not long been on the way, but it was the pandemic that endowed what had formerly been contestable with an aura of inevitability and common sense. Most dramatic, for me, was the new public health regime that installed itself in March 2020 almost without comment in the media. Richard Schabas summed it up very concisely in testimony to a citizen's inquiry that was convened by the Canadian COVID Care Alliance in June of 2022.[20] Throughout his long career, he said, public health had been supposed to rest on four principles: the priority of the public interest as a whole; the primacy of evidence; a preference for persuasion over coercion; and the minimization of fear and panic. All these principles, he said, went out the window in March 2020. A "war" against a single disease was pursued at all costs; draconian measures, like lockdown, were implemented without evidence of their effectiveness or concern about their side effects; vaccination, distancing, and masking were made mandatory; and panic was heightened rather than damped by dissemination of speculative, worst-case models. It seems legitimate to call such a complete reversal a revolution, and this was exactly what the public health veterans who spoke out in the summer of 2020 quietly tried to say had happened, but the magnitude of the change seemed to pass unnoticed outside these circles.

The CBC ought to have been the place where such things *were* noticed, the place where the narrative in which the pandemic arrived, as it were, prepackaged should have been questioned and investigated. Instead, it functioned as a space of acclamation, rather than of questioning. The cautions of senior, impeccably credentialed Canadian public health authorities were brushed aside. This seems to me to contradict the CBC's most basic *raison d'être* as a public broadcaster—to provide an open and uncommitted space in which questions can be raised and differences explored. Questionable policies were adopted, questionable assumptions were made—but no questions were asked.

I view this as an evacuation of the very purpose of public broadcasting. For sixty years, as I have shown, the CBC has been trying to get closer to its audiences and to assemble those audiences into a homogeneous and unanimous "we." I believe that the pandemic shows where this approach finally leads. The CBC has become a captive of the very unanimity it has worked so hard to construct. Questions that demand to be patiently and dispassionately opened are instead foreclosed. The difficult work of gaining distance from puzzling phenomena is forgone in favour of the close embrace of the taken-for-granted. Those who put themselves outside this consensus are assigned derogatory names and disqualifying labels. At the very moment when Canada needs a free space for critical thought, the CBC has opted for the view that the truth is already securely in its possession and need only be preached to the choir.

PART TWO

7 | What Is a Public?

A new vision for public broadcasting will depend, first of all, on a renewed understanding of what the public is. In my Introduction, I wrote of "the collapse of modern certainties," the breakdown of the taken-for-granted, consensual understanding on which the CBC was founded and within which it has operated. The existence of *the public* is just such a certainty. Within my lifetime, one could still speak confidently of *public* broadcasting or the *public* interest, and expect to be understood without question. I don't think this is any longer the case. One can, of course, still speak in this way and even hope to gain a certain legitimacy by using this once-exalted word, but the term is now apt to produce confusion and controversy rather than clarity. How the public is constituted, and who constitutes it, are questions that have grown obscure and need to be rethought.

The primary split is between those who imagine the public as a compact or unitary body, and those who see it rather as a dynamic field comprised of multiple *publics*. Populism at the CBC has fallen squarely into the first camp. From Ross McLean's "How will this serve *the* audience?" to Richard Stursberg's "coherent brand promise," the CBC of the second era has addressed the public as if it were all of one ideological tenor and one moral consistency. This supposed homogeneity no longer exists, and perhaps never entirely did. But, whether or not the seams at which the public is now being pulled apart were there all along, it is certainly true that the public no longer feels itself to be as one. In places where the divisions are most acute, as in the United States, predictions of impending civil war are not uncommon.[1] Reconceiving the public as a series of distinct publics, each one standing on different premises, will not create instant peace, but it will at least open a potential pathway to a forum where peace could be negotiated.

Our idea of the public, by that name, traces back to the eighteenth century. Consider, for example, a discussion that took place in the French National Assembly in 1791 concerning the constitutional significance of public opinion.

One of the delegates, Nicolas Bergasse, told the assembly: "You know that it is only through *public opinion* that you can acquire any power to promote the good; you know that is only through *public opinion* that the cause of the people—for so long give up as hopeless—has prevailed; you know that before *public opinion* all authorities become silent, all prejudices disappear, all particular interests are effaced."[2] A year later, in England, Charles James Fox, the leader of the Whig opposition, made a comparable statement in the House of Commons. "It is certainly right and prudent," he said, "to consult *the public opinion*. If *the public opinion* did not happen to square with mine ... I should consider it as my due to my king, due to my country, due to my honour to retire that they might pursue the plan which they thought better."[3] In both statements, *the public opinion* is conceived as a single, sovereign, and palpably real entity. In the United States, whose constitution was written around this time, the whole document is imagined as the utterance not of its authors but of a collective being called "We, the people ..."

A hundred years before none of these declarations would have made any sense. In 1672, for example, the English king, Charles II, issued a proclamation commanding his "loving subjects" not to "intermeddle with the affairs of state and government." A second proclamation, two years later, condemned "licentious talking of state and government." It's true, of course, that Charles would not have issued these proclamations had he not feared that an emergent public was indeed beginning to "intermeddle." But he was also reiterating the still prevailing theory, according to which publicity, or publicness, was entirely the property of those whose rank made them public persons. The king *was* England; the Duke of Cornwall *was* Cornwall. You can hear it still in Shakespeare: "Here's France," "Speak, Kent," "Farewell, Gloucester," and so on. Places and polities spoke through their noble embodiments. "Matters of State," Queen Elizabeth proclaimed to her subjects in 1559, were fit to be treated only by "men of authority" and conveyed only to audiences of "grave and discreet persons." Publicness was inherent in those who by birth or delegation possessed it—it was something enacted *before* the people, not derived *from* the people. Nothing could become public that was not made public by someone entitled to represent the realm in this fashion.

So how did private, non-entitled persons gain the capacity to make things public? The classic account comes from German philosopher Jurgen Habermas in his book, *The Structural Transformation of the Public Sphere: An Inquiry into a Category of Bourgeois Society*. Habermas argues that during the course of the eighteenth century, something new came into existence. A civic space began to take shape outside the church and the state and began to claim the name that had previously belonged exclusively to the rulers of church and state: the public. Private individuals began to associate in what Habermas calls their "common quality" as communicative beings, and the opinions they formed began to count. The conditions that allowed this new public sphere to emerge, he wrote, had been maturing ever since the Reformation. Notable among the factors he emphasized were the achievement of the freedom of religion; the establishment of a domestic sphere in which a new kind of solitude, subjectivity, and intimacy became possible; the expansion and increasing freedom of the press; and the emergence of critical audiences for various arts including theatre, painting, and literature. Coffee houses and salons provided new venues for conversation; magazines and journals provided new vehicles by which, "Philosophy [was] brought out of Closets and Libraries … [and made] to dwell in Clubs and Assemblies," as Joseph Addison wrote in *The Spectator* in 1711.[4] By the end of the century, this new thing was strong enough, in Habermas's words, to "compel public authority to legitimate itself before public opinion."[5]

Jurgen Habermas's picture of the bourgeois public sphere was clearly an idealization, as many critics pointed out.[6] Habermas played up "the constant, discreet use of … reason" that he observed in the world of polite letters and urbane conversation, and tended to overlook evidence that the eighteenth-century public sphere, like our own, was, equally, a theatre of passions.[7] He also recognized, in the second half of his book, that technological and commercial developments soon put an end to the free creation of public opinion that he thought had at first characterized the bourgeois public sphere. Indeed, he is quite dire about the "decay" and "disintegration" that he felt marked the public sphere from the nineteenth century onward.[8] Beginning with the first stirrings of mass media, he wrote, "trust in the power of reason … shatter[ed]," and public opinion began to be "manufactured."[9] Where once a genuine "domain of interiority" had been "oriented to a public audience," now the "patterned … predigested" and manipulative offerings of mass media invaded and disrupted "the inner life."[10] "In the world of letters," he wrote, "… the public sphere was replaced by the pseudo-public or sham private world of culture consumption."[11] "The law of the market," which once governed only the distribution of cultural commodities, "ha[d now] penetrated into the substance of the works themselves and … become inherent in them as formative laws."[12] The public sphere, in short, had been "refeudalized," and the old "representative publicity" of the nobility reinstituted in the contemporary cult of celebrity.[13]

Habermas's indictment of the mass public echoes criticisms that were first articulated in the nineteenth century by liberal theorists like John Stuart Mill. In his *On Liberty*, Mill treated public opinion, not as the sum and quintessence of rational deliberation, but as a form of "moral coercion" that was "intolerant of any marked demonstration of individuality" and apt to "make people all alike."[14] His French contemporary Alexis de Tocqueville expressed similar concerns. "[Public opinion] uses no persuasion to forward its beliefs," Tocqueville wrote, "but by some mighty pressure of the mind of all upon the intelligences of each it imposes its ideas and makes them penetrate into men's very souls."[15] The Danish philosopher Søren Kierkegaard, writing at the same time, was even more scathing. The public, he said, is nothing more than "a phantom"—"an abstract and deserted void" in which the real existence of the individual person is "levelled" and lost.[16] "In order that everything should be reduced to the same level," Kierkegaard writes, "it is first of all necessary to procure a phantom, a spirit, a monstrous abstraction, an all-embracing something which is nothing, a mirage—and that phantom is *the public*."[17]

Since these opinions were expressed, the contrast between Habermas's ideal, rational public and Kierkegaard's "deserted void" has been drawn again and again. The mass public has been condemned as an intolerant and dominated dupe; the democratic public has been glorified as a wise sovereign. The two views seem irreconcilable so long as the thing at issue is seen to be *the public*—a compact and consistent being able to write constitutions, silence authorities, and arrive at rational judgments. Disaggregate the public, and the problem suddenly becomes more manageable. Behind the mirage of *the public*, multiple *publics* begin to appear. This was a point made a hundred years ago by journalist Walter Lippmann in a book called, notably, *The Phantom Public*. By personifying abstractions, like *society*, he wrote, we create unthinkable and intractable problems. But these problems are:

> … not so insoluble once we cease to personify society. It is only when we are compelled to personify society that we are puzzled as to how many separate organic individuals can be united in one homogeneous organic individual. This logical underbrush is cleared away if we think of society not as the name of a thing but as the name of *all the adjustments between individuals and their things*. [my italics] Then we can say without theoretical qualms what common sense plainly tells us: it is individuals who act not society … it is the painters who paint not the artistic spirit of the age; it is the soldiers who fight and are killed, not the nation … It is their relations with each other that constitutes society. And it is about the ordering of their relations that the individuals not executively concerned in a specific disorder may have public opinion and may intervene as a public.[18]

Lippmann's point that publics are partial, practical, and pointed came home to me through my participation in a five-year research project called *Making*

Publics, which I covered for *Ideas* between 2003 and 2008.[19] This multi-country academic consortium, centred on McGill University, undertook the task of unearthing the history of what Habermas amalgamated under the name of the public sphere.[20] What they found was that long before anyone spoke of *the public*, there were *publics*—active and interested assemblies that came together around matters of concern to them. Church historian Torrance Kirby, for example, found that a "first instance of a public sphere" took shape around the religious controversies of the Reformation period.[21] The Church had split, and the once unanimous and enforceable truth of religion had come into question. What Walter Lippmann nicely called "adjustments" were necessary. Could an earthly ruler really be the supreme head of the Church in England, as Henry VIII had proclaimed himself to be? The crowds that assembled at the outdoor pulpit in the churchyard of St. Paul's Cathedral had to be persuaded that he could. The same principle was repeated in scores of other fields. The printing press gathered publics around the interpretations of mass-produced texts. Painters and playwrights convened publics in discussion of the images and ideas that they proposed. Writer and scholars assembled publics in support of their movement to replace Latin with their national languages. Affinities created new forms of association in the same way as matters of concern. Mapmakers and madrigal singers, antiquaries and astronomers formed new societies of interested amateurs.

Out of these many instances a pattern emerged. A public is the expression of some vital interest that draws people together. It's a vortex of attention—open at its edges—not yet amounting to an institution—but still exerting a pull on the shape of the surrounding society. Publics are connected by the things that matter to them. They may be physical things like maps, printed broadsides, or paintings, or they may be ideas like the Reformation of the Church or the cultivation of the English language. Publics are virtual—they require always a certain imaginative reach. A reader, alone, imagines other readers. A radio broadcaster imagines listeners and hopes to assemble a public, but never entirely knows to what extent that has happened. A spectator at a sporting event imagines that it's his city or his country that is playing. And publics, finally, are creatures of media—the means by which things get around and a common awareness is made possible. Communications media are crucial, but they do not exhaust the idea, for paintings or landscapes or printed musical scores can be media too.

Society in this account is something that is continuously being made, and made, so to speak, on the fly. It is not an empty container that people fill up with facts, fables, and controversies of all kinds—it is the facts, fables, and controversies that make society in the first place. So, attending to the ebb and flow of publics yields a different picture of society than the study of society's more rigid and durable institutional features—a more fluid picture of the underlying movements of interest, affinity, and desire. It also yields what is in some ways a more encouraging image. A consolidated public appears inert, incompetent, and easily manipulated—like the hive mind of Tocqueville's nightmares or Kierkegaard's bemused citizen consumed by the "forest fire of abstraction."[22] Active publics, engaged in practical adjustments, assembled by definite and limited interests, escape this stricture. But to thrive they must first be recognized and allowed their proper scope.

The Modern Constitution

This is particularly important at a moment like ours when crucial certainties have collapsed, historical landmarks have disappeared, and people have divided, politically and culturally, into hostile camps inspired by incommensurable worldviews. The modern age once possessed a distinct constitution—a set of guiding ideas with inspiring names like progress and development, democracy and equality, society and nature. French philosopher of science Bruno Latour described this constitution in his pathbreaking book *We Have Never Been Modern*, published in English in 1993. His book, along with several of its sequels, has had such a formative influence on my thinking about publics, that I want to explore its argument briefly before returning to my main theme. The modern constitution, Latour wrote, rested on a series of clarifying and simplifying separations, or, as he said, "purifications." The crucial one was the drawing apart of nature and society. God as an active agency was "crossed out" and replaced by a deified nature. Science became the handmaiden, priest, and oracle of this new god. The body politic was set apart from nature and defined as consisting entirely of people and their opinions. Subjects and objects were segregated. Enlightened thought had no worse term of abuse than "naturalization," or finding in the object what properly pertained only to the subject. Society was treated as artificial, something made by people, while nature was taken—there's no other way to say it—as natural. By this strict division, nature was opened to unlimited exploitation, as all considerations of its inherent meaning had been assigned to the sphere of opinion—nature was comprised of facts, society of values. Science became nature's voice, and, in this way was seen as standing apart from and opposed to politics.

This ideal scheme and its large supporting cast of purified pairs—nature/culture, object/subject, fact/value, etc.—now lies in pieces. What has undone it, according to Latour is the appearance of what he calls "hybrids"—entangled objects that straddle these old dichotomies, objects that participate equally in nature and society, and so absolutely resist purification. What sort of thing, for example, is a genetically engineered organism? How about a vaccine that "reprograms" your immune system? Are they nature or culture, subject or object, fact or value? Questions about how nature is represented have become inseparable from questions about how the human political community is represented.

Climate change, which now sits at the very centre of our politics, is just such an entangled object. Earth's atmosphere is plainly a natural thing, but just as plainly an "anthropogenic" one, according to the discourses now driving public policy. But how do we know that human actions are responsible for a changing climate? Weather is perceptible; the climate is not. The climate is an elaborate and contingent scientific construction—the creation, in Latour's words, of "a multitude of earth sciences whose certainties have been achieved not by some earth-shaking discovery but by the weaving together of thousands of tiny facts, reworked through modelling into a tissue of proofs that draw their robustness from the multiplicity of data, each piece of which remains obviously fragile."[23] Latour believed that climate change was of human origin. Indeed he believed that it marked a decisive turning point in human history. But he still recognized that the science of entangled objects could never achieve the certainty once promised by the sciences of stand-alone nature. He knew that no amount of bravado about "settled science" could imbue millions of measurements coaxed together into unimaginably complex simulations with unchallengeable authority.

A new situation therefore exists and one quite outside the clear parameters of the modern constitution. Modernity was a world of clear demarcations. We now live in a world of fuzzy objects with uncertain boundaries—a world of unforeseeable, incalculable, and indefinable risks—a world quite the opposite of the controlled and coordinated space that modernity had supposed itself to be. Nature and society have become indistinguishable; science has become contestable; and a clamorous crowd of new actors, ranging in scale from antibiotic-resistant microbes to thermonuclear weapons, have entered the political scene. With its products overwhelming society, technoscience can no longer stand aside as the disinterested voice of nature.[24]

A situation in which no one is quite sure what is what, and who is who, demands a redefinition of public space. According to Latour, modernity knew no true politics, because the modern constitution settled all major questions in advance. Science told us what the world was made of; progress set its direction; differences of opinion assorted themselves as variations within this authoritative framework. Now we face real questions without authoritative answers—questions on which people can and do legitimately differ. Politics, therefore, can "begin again," and must begin again, if social peace is to be restored.[25] Currently, people are divided on moral grounds—they disagree about what is good—on philosophical grounds—they disagree about what exists—and on political grounds—they disagree about what properly belongs to whom. These differences, as I've said, are incommensurable—there is no common measure to which they can be reduced, no critical analysis that can clear everything up and get everyone re-aligned. No peace will be possible, until we first recognize "that humans are divided into so many war parties."[26]

The enmity of these war parties can be overcome, according to Latour, only by our accepting a "diplomatic obligation to introduce ourselves to one another in the form of newly defined peoples"[27] What Latour calls *peoples* here might refer to a territorial people, but also to various other subcultural identifications. He uses the term "collectives" more or less interchangeably. The key idea is that such groups must understand the full extent of the otherness of the others who confront them. This will entail people asking each other fundamental questions about their "supreme authorities," about their conception of the land on which they live, and about the "time period [in which] they situate themselves."[28] Diplomacy, peacemaking, and conversation can only proceed when an existing state of opposition and enmity is recognized. There is "no politics without real opponents."[29] This condition is not met where opposing parties scorn, vilify, and disqualify one another. Diplomatic negotiations cannot begin where opponents recognize no legitimate differences of opinion and instead ascribe all disagreements to misinformation, bloody-mindedness, or bad character.

To summarize rather bluntly: We're now in a situation in which no one knows what they're talking about, and everybody thinks they're right. The science that was supposed to quieten and overrule dangerous political passions now provokes them instead. History is on trial. Progress has lost its way. Public speech has become a minefield. This social, intellectual, and spiritual shambles calls, insistently, for a new conception of the public, a conception which must be above all plural, catholic, and open. Let the idea of *the public* by all means linger—as the distant hope of unity—but what must be first of all acknowledged is that down on the ground the public is comprised of many publics that are constantly assembling and disassembling themselves. Publics, as I have tried to show, are spontaneous and changeable, with blurred boundaries and shifting casts of characters, possessing no more stability than a standing wave in flowing water. This mutability makes them a *means* of change and adjustment as well as an *expression* of change. Publics bring matters of concern into general view. They allow strangers to assemble and find one another.

They give private concerns and affections public voice, and, in this sense, they are both personal and impersonal at once, allowing the personal to become political without the political becoming destructively personal.[30]

A change in how the CBC imagines civic space—a turn to the plural and the complex—would open a new path for public broadcasting. The CBC, during its populist era, acted *for* its public, substituting itself for its audiences as their delegate, clown, and avenging angel. It was *your CBC* where *you* could find laughter and inspiration, justice, and vindication. Canada was never really as homogeneous as this myth asserted, but the story, in its time, aligned well with many other features of post-1960s Canada. It "worked." Now this image has imploded. Humpty Dumpty has had a great fall and cannot be put back together again. Invocation of *the public* is now nothing more than a veil hiding the CBC's ideological prejudices and sheltering its preferred styles of thought. *Publics*, on the other hand, are active, exigent, demanding creatures. They are, in Hannah Arendt's term, "world-making"—the opposite of the passive resource that *the public* has now become. To learn to see itself as an interlocutor of active publics, rather than as the voice of a like-minded community, the CBC will have to change. It will have to allow the deep divisions in Canadian society to emerge and become articulate. It will have to liberate the power of publics to bring matters of concern into focus. And it will have to become the public forum that Canada will desperately need in the uncertain new age we are now facing.

The Freedom Convoy as a Public

I'll conclude with an example that I think shows how things currently stand and why we urgently need the new thinking about how the public is constituted that I've been recommending. In January 2022, in the dead of winter, a convoy of trucks originating in several parts of the country began to converge on Ottawa, gathering strength and support as it rolled toward the capital. Styling itself "The Freedom Convoy," the protest tied up the city for several weeks before being broken up and dispersed by the police in late February. This event was spontaneous and leaderless, with its various tributary streams coming together out of a shared disquiet rather than some prior design. There were organizers, of course, and precedents, but planning was sketchy, and provision was mostly impromptu.[31] Even the people subsequently identified as "leaders" had not all known each other in advance. Along the way, crowds of strangers assembled on bridges and in malls and parking lots to express their support. People were "finding one another," as one person who was there said.[32] What brought them together was a shared matter of concern—mandatory vaccination policies. Some were worried about how vaccine mandates had changed the character of the country. They saw a harbinger of tyranny in a government that would force a novel medical treatment on its citizens without debate. Many others had an even more personal stake—they had lost their livelihoods and access to many formerly public places when vaccination became a condition of employment and public participation.

Watching this event take shape, I was astonished at both the initiative and the depth of feeling that were evident in this surprising undertaking. With temperatures at −30, no guarantee of financial support, and people's very livelihoods sometimes at stake, getting and keeping those big rigs rolling seemed to me an impressive display of faith and courage, as well as of political feeling. I even remember briefly hoping, as the convoy gained momentum and popular support, that its size, obvious verve, and sheer improbability might gain it the government's ear and start a national conversation.

But the government and the respectable media accorded this assembly no legitimacy whatsoever. According to the prime minister's top national security adviser, in her testimony to the subsequent public inquiry, these were "violent extremists" bent on "overthrowing" the government.[33] The CBC set the tone for its consistently hostile coverage on the weekend the trucks arrived, when the nightly news broadcast *The National* chose to interview a trucker who was against the protest rather than one of the participants. The same attitude was evident when *The Fifth Estate* tried to sum up what had happened several weeks later.[34] In Gillian Findlay's full episode documentary report, the issue that had brought the protestors together was barely mentioned. Instead, the convoy was treated entirely as a problem. It was discussed as a failure of intelligence and policing and as a manifestation of "extremism." The bad manners and questionable opinions of its "leaders" were challenged. Through the *Fifth Estate's* lens, everything about the protest seemed calculated and slightly sinister, even down to the confession by James Bauder, one of the convoy's first promoters, that the idea had come to him "in prayer." When the demonstrators' opposition to vaccine mandates came up at all, it was under the heading of misinformation, since that was the only reason for being against compulsory vaccination that the CBC recognized.

This unwillingness to take the convoy and its supporters at their word continued in subsequent commentary. Take, for example, an op-ed in *The Globe and Mail* by pollster Frank Graves and veteran journalist Michael Valpy, headlined, "To avoid future 'freedom convoy' protests, we need an economy based on hope."[35] Graves and Valpy begin by declaring roundly that "the supposition" that the occupation of Ottawa was "primarily driven" by the issues that the protestors said animated them—mask and vaccine mandates—is "flat out wrong." These were mere "surface-level flashpoints and rallying cries," they say. Underlying these ostensible and explicit causes, our writers discern three "hugely significant" trends: authoritarian populism, loss of "institutional trust," and social media-generated disinformation. And, behind these in turn, they diagnose diminished opportunity and loss of status for "males under the age of fifty who are lacking university educations."

These explanations may well point to something that is real, but observe how the account the protagonists in this story gave of themselves has completely disappeared. What the people who came to Ottawa said they were doing, and apparently thought they were doing, was presenting an argument. This was, in brief: that lockdowns had done more harm than good; and that a hastily tested vaccine employing a novel genetic delivery system ought not to have been imposed on people at the cost of their livelihoods and civil rights, even if it had been effective in preventing transmission of the virus, which, in the event, it was not. Both propositions are, at the least, arguable on both moral and scientific grounds. Lockdowns *did* produce harms—harms whose full extent have yet to be appraised because these harms are usually ascribed to the virus itself and not to the policy response. The summary abolition of the principle of informed consent, proclaimed at Nuremberg in 1947, *did* overturn what had previously been said to be the very foundation of contemporary medicine's legitimacy. But both these propositions were dismissed as mere froth, and the protestors' views were said to be the product of misinformation. Everything tangible and intelligible in the protest was set aside, and what the people who were there said they were concerned about was replaced by expert diagnosis—these were ignorant and angry people expressing resentments that their lack of university education made them unable to recognize.

What the politicians, the media, and the pop sociologists failed to recognize, in my opinion, was an emergent public. Suddenly and unpredictably, a large group of strangers assembled around an issue of common concern. Anyone who felt summoned, or addressed by this issue was eligible to join. Boundaries were fluid and blurred, leadership was mostly improvised and sometimes contested. The meaning of what was occurring was up for grabs. People were doing a new thing in response to new circumstances. Their ideological predispositions, and their socio-economic circumstances, were relevant, of course, but these frames did not exhaust, and in some ways did not even comprehend, what was going on. What manifested itself on Parliament Hill, and the roads leading to it, was a dynamic chaos, a potentially creative disorder seeking form and interpretation as it went along.

The protest was certainly inconvenient, unruly, and sometimes uncouth—F🍁CK Trudeau being one of the protestors' more prominent slogans. All those throbbing semis understandably unnerved Ottawa residents, as did the blaring air horns, until a court injunction finally silenced them. But the demonstration was also very largely peaceful and friendly, according to everyone I talked to who was there, and, for many of the participants, offered a life-altering experience of political community.[36] To me, it looked like an eruption from what might be called Canada's political unconscious—something unforeseen pushing into awareness, a sign that people had been pushed too far and were now, almost reflexively, pushing back. The government had imposed a vaccine mandate on the truckers who had delivered the goods, while the laptop class comfortably "sheltered in place," and had done so at a time when this trucker group was already nearly 90 percent vaccinated, and it was already clear that vaccination was not preventing transmission in any case. Something just snapped. What it would come to *mean* was undefined and undetermined.

This open and mutable character was almost entirely overlooked by those with the power to fix a meaning on the convoy and its supporters. The government saw only nuisance, threat, and the manifestation of what the prime minister called "unacceptable" opinion. Indeed, the prime minister quickly surmised that "these people" were "often ... racist ... misogynist ... and anti-science."[37] The media, including the CBC, employed ready-made frames like misinformation, conspiracy, anti-vaxx, etc., to avoid any consideration of what the demonstrators were trying to say. The pundits, like Valpy and Graves, felt no need to listen, because they already understood the protestors better than they understood themselves. Whatever interpretation resulted—security risk, victims of the internet, populist resentment—little account was taken either of the remarkable gumption the truckers and their allies showed in putting themselves and their livelihoods on the line in the dead of winter, or of the conversational character of the proposition they were presenting in taking their stand on the issue of freedom. Nobody wanted to talk, and nobody wanted to engage with the significance of what was taking place.

The trucker convoy was summoning and assembling a new public, but this new formation went for the most part unrecognized. I believe that the reasons for this lack of recognition go beyond the trouble the convoy created. I don't want to downplay this disruption—chaos precedes creation, and bringing a new thing into being requires a potentially violent energy that is not easily contained—but I also think that this new public went unrecognized and unaddressed because those whose recognition might have made a difference do not know what a public is. For them, *the* public is essentially an inert idea. It is a mass—of voters, consumers, listeners, etc.—whose vocation is to be acted *on*. The prime minister's word *unacceptable* sums up the case. What was appearing was something he and his counterparts in the media and the cognoscenti didn't want to see, so they ruled it out of bounds, and felt it as their right to do so, the public being essentially theirs to define and construct according to their own lights.

This has created what I see as an ominous situation. If the dignity of an emergent *public* had been granted to the convoy, conversation and accommodation would have had to follow. In the event, what occurred was ostracization, projection, and all the other dynamics of enemy-making. The convoy was excluded from the public. This is unlikely to end well. If a rigid stereotype is all the space that is granted to people who try to start a conversation, then they will tend to grow into that stereotype, adopting the role that has been written for them. Tell someone that they are misinformed, without recognizing that you might be misinformed yourself, and that person will likely retrench on the misinformation that you think defines them. What I have hoped to do in this chapter is to frame a definition of the public that might make it possible, in future, to bring manifestations like the Freedom Convoy into conversation rather than driving them to the margins of public life.

This should concern the CBC as a *public* broadcaster above all others. One can understand a government that takes a protest with F🍁CK Trudeau emblazoned on its standard as hostile. One can even understand the *déformation professionelle* of intellectuals who would rather explain than understand. But the CBC is specifically charged with the maintenance of a public forum in which Canadians can speak to each other and hope to be heard. If it does not model openness and civility, who will?

8 | Beside Ourselves: A Meditation on Media

"... the basic error in modern times is just this: being continually concerned with what one needs to communicate rather than with the nature of communication."

—*Søren Kierkegaard*[1]

The goal of science and the arts and of education for the next generation must be to decipher not the genetic but the perceptual code. In a global information environment the old pattern of education in answer-finding is of no avail; one is surrounded by answers, millions of them moving and mutating at electric speed. As the information that constitutes the environment is perpetually in flux, so the need is not for fixed concepts but rather for the ancient skill of reading that book, for navigating through an ever-uncharted and unchartable milieu. Else we will have no more control of the technology and environment than we have of the wind and the tides.

—*Eric and Marshall McLuhan,* The Laws of Media: The New Science[2]

In the preface to his final work, *The Bias of Communication,* Harold Innis wrote that his book was an attempt to answer "an essay question ... which the late James Ten Brooke, Professor of Philosophy at McMaster University, was accustomed to set." The question was: "Why do we attend to the things to which we attend?"[3] Innis had been Ten Brooke's student in the years before the First World War, and the question, he said, had preoccupied him ever since. *The Bias of Communication* was his answer: We attend to the things to which the available media of communication direct our attention.

This was a revolutionary thought, but one that has so far proved difficult for the practitioners of media to think. Their primary question, understandably, has always been: What can be done with a given medium, not what does that medium itself do? Caring for the effects of media, consequently, has barely been noticed as a task. I think this stance now ought to change in view of the increasing media intensity of contemporary society.

Philosopher Søren Kierkegaard, writing in the 1840s, claimed that "the basic error in modern times" had been a preoccupation with "what needs to be communicated" at the expense of "the nature of communication." Marshall McLuhan would later formulate the same distinction as the difference between the message and the medium. Various other pairs exemplify an analogous contrast: content and container, means and end, word and language, conscious and unconscious structure, figure and ground, and so on. However the distinction is framed, the emphasis is apt to fall on what is in the foreground, while the infrastructure by which it has been made to appear recedes into the background and soon escapes our notice altogether. We look past the means to the desired end.

This attitude is viable only so long as a society and its media remain aligned and in harmony. Print literacy, to take one of McLuhan's examples, was "the technology of individualism."[4] As such, it aligned with, and supported, many congruent tendencies within Western society and, for that reason, its character as a medium tended to remain out of sight and out of mind. Today, the situation is dramatically different. We are inundated by communications media, and their effects are often strikingly misaligned with the ostensible ideals of our society. At times, it appears that the world itself is dissolving in its representations, thus creating the widely reported crisis of truth, or even, more drastically, of reality.[5] Under these new circumstances, a purely instrumental attitude, interested only in exploiting media while overlooking their by-effects, seems imprudent, to say the least. Accordingly, I will devote this chapter to the question of what media are doing *to* us, rather than *for* us, and to the question that follows from it: Can public broadcasting find a way out of the hall of mirrors that the proliferation of communications media has produced?

* * *

The word *medium* is expansive and can denote almost anything that comes between two other things as a mode of connection or carriage. Earth and air, fire and water are all media.[6] But the word in the sense that concerns us here traces back to Canadian economist Harold Adams Innis. Innis set out in the 1920s to explore the economic pathways by which Canada became Canada. He published his findings in books like *The Fur Trade* and *The Cod Fisheries* and various other studies of Canada's resource economy. Then, in the decade before his death in 1952, he decided to "test the limits of his tools" by bringing "the bias of an economist" to bear on the phenomenon of communication. He concluded that the very "character of knowledge," right down to the ways in which we knit the elements of space and time together into an intelligible world, depends on the "media of communication" available to us. In oral societies, Innis wrote, the evanescence of the spoken word requires an emphasis on continuity in time—on memory, custom, and the preservation of tradition. Writing, on the other hand, permits projection in space, which increases as the media of inscription grow lighter and more portable. "Monopolies of knowledge" are built up by those who control the means of communication.[7]

Innis's research was highly original. Marshall McLuhan may have been extravagant in saying that his own work was only a footnote to Innis, but it's certainly true that Innis opened the path that scholars like McLuhan, Walter Ong, Neil Postman, and many others would later explore.[8] There is another sense, however, in which Innis was only the Canadian outcropping of something much larger: the unsettling discovery of *media*tion that was being made across all fields in the first half of the twentieth century. At the University of Toronto, Innis's friend and interlocutor, Eric Havelock, was studying the ways in which alphabetic literacy transformed consciousness in ancient Athens. Academic philosophy was taking the so-called linguistic turn—the discovery that thinking cannot transcend the linguistic medium in which it is transacted. In anthropology, Benjamin Lee Whorf noted that the features of the world that will be most salient for a given culture will be those brought to light by its language.[9] In psychology, the positing of an unconscious mind led to the idea, in Carl Jung's words, "that ... the subject of knowledge, the psyche [, is] a veiled form of existence not immediately accessible to consciousness" and, therefore, that "all our knowledge must be incomplete ... to a degree that we cannot determine."[10] In science studies, Ludwik Fleck pointed to the dependence of scientific facts on what he called "thought collectives"—ensembles that laboriously work out the definitions and agreements by which facts are stabilized, recognized, and sustained.[11] Thomas Kuhn would later call the master structures that make facts intelligible "paradigms," and, since Kuhn wrote, historians and philosophers of science have continued to trace the networks, thought styles, and physical infrastructure underlying scientific knowledge.

In all the cases I have mentioned, the characteristic discovery has been that all knowledge is mediated, in the largest sense of that term. We perceive "through a glass darkly" and the defects and distortions of the glass become part of what Innis calls the "unevenness" of our thought.[12] All human knowledge is inherently dependent, partial, and flawed. You cannot see the means by which you are looking. It follows, says McLuhan, that we often walk "the beaten paths of impercipience," unaware of our unawareness.[13]

Media Apocalypse

McLuhan's conclusion is undeniable, but not, I hope, inevitable, since I can see no way that we can live well in the media-saturated world that has already come to pass without our bringing into awareness the so-far mainly imperceptible grammar of media. A note of apocalyptic urgency, in this regard, was already sounding in the last writings of Harold Innis. Innis died in 1952, just two months after CBC Television began broadcasting, at a time when radio and the mass circulation newspaper were still the dominant elements of the media environment, but he already perceived a crippling dislocation and disorientation of sensibility. "The cruelty of mechanized communication" had produced, he said, "a ruthless shattering of language," and a "systematic destruction of the elements of permanence essential to culture."[14] Industrial and electric media, which emphasize projection in space at the expense of persistence in time, were beginning to lead into a "present-mindedness" so total as to threaten a "paralysis" of thought itself.[15] Time was being cut into ever more "precise fragments suitable to the engineer and the accountant."[16] The controlling eye held the listening ear in thrall, Innis thought, and all sense of the value of duration, tradition, and conversation across time was threatened as a consequence.

Innis, by his own account, was a wounded man. Right into his last years, he described himself as a "psychological casualty" of the First World War, and he was certainly more acutely sensitive than most to the technological violence of his civilization.[17] Even so, what I would call the apocalyptic logic of his late writings seems to me entirely prescient. Colloquially, the word *apocalypse* has become a synonym for disaster, but I mean it here in its original ancient Greek sense of revelation. By the logic of apocalypse, things press toward the point at which their nature is fully revealed. "You never know what is enough until you know what is more than enough," says poet William Blake in his aptly named *Proverbs of Hell*.[18]

The sheer glut of media forces itself on our attention. McLuhan, writing at the end of his life, predicted that an unexamined media environment would acquire an inexorable power equivalent to that of "the wind and the tides." Many now echo him unconsciously in speaking of a "media ecosystem"—as if the artificial environment had already become a natural environment. An example of this way of speaking would be Douglas Rushkoff's *Media Virus!*

(1996), a book entirely taken up with proposing artful adaptations to the quasi-natural environment created by media. But I do not believe that we can simply adapt to this environment—however agile the *jujutsu* by which the "media-savvy" try to turn it to their own benevolent purposes. I think we need to understand, first of all, what this environment is doing to us. Only then will it be possible to find out the ways in which we can limit, steward, and use these media.

One of the writers who has helped me to understand the situation we are in is French philosopher and critic Jean Baudrillard (1929–2007). Baudrillard's essential idea is that highly mediated contemporary societies have crossed a threshold into a condition that he calls a *simulacrum*—a condition in which reality is, in effect, beside itself, overwhelmed by its representations to the point where reality and representation have become indistinguishable. At this threshold, media begin to generate reality rather than mirroring it. Models become self-fulfilling prophecies, functioning, Baudrillard says, as "magnetic fields" that pull events into the shapes that have been prepared and predicted for them.[19] Maps begin to "engender the territory."[20] Photographs replace description, stereotypes drawn from pop sociology and psychology supplant analysis, verbal formulas substitute for considered speech, and typifications of all kinds absorb the unique and unrepeatable. Signs evoke other signs, like the Vedic image of Indra's net, in which every knot is a multi-faceted jewel, and each one reflects all the others without origin or end.

Baudrillard names this new state "hyper-reality"—a condition in which the represented is "more real than the real."[21] Hyper-reality, he says, has two catastrophic entailments. The first is that it provides no grounds for moral or aesthetic judgment, being itself groundless. The second is that it throws events out of any stable or meaningful sequence. At "the speed of information," he writes, events "volatize," and history has no time to take place.[22] Events acquire "the strange aftertaste of something that has already happened before, something unfolding retrospectively … our only surprise is that we were not able to foresee them and our only regret that we don't know how to draw any consequences from them."[23] History "curves back" on itself, and this imparts to the affected societies what Baudrillard calls a "revisionistic" character. The will to make history is replaced by the will to unmake it. "Our societies," he writes, "… are quietly rethinking everything, laundering their political crimes, their scandals, licking their wounds …"[24] Things happen, but with no time in which to take place, their novelty and promise are quickly contained, and they are pulled back into the gravitational field of an endlessly present past. Simultaneity smothers history, even as history continues, leading to a sense of an ending that never ends—an apocalypse continually deferred.

Baudrillard's insights into the consciousness of excessively mediated societies can be unnerving, uncanny, and occasionally, overwrought. Yet his observations match those of many other contemporary thinkers who've been haunted by the idea of a world swallowed by its own image. Daniel Boorstin, in his 1961 book *The Image*, foresaw the coming of the "pseudo-event"—an event that is, so to speak, pure publicity and possesses no reality outside its own staging. Milan Kundera's popular novel *The Incredible Lightness of Being* (1985) alluded to an experience of weightlessness and inconsequence that is very similar to what Baudrillard tried to convey. American writer George W. S. Trow, in his *Within the Context of No Context* (1981), narrates the transition from "history [as] the record of growth, conflict and destruction" to "history [as] the record of demographically significant preferences"—a "New History [in which] nothing was judged, only counted."[25]

The basic premonition, in all these authors, is that personal action has given way to managed behaviour, personal judgment to definitive expertise, moral discernment to correct categorization, and intelligible sequence to an endlessly repeated present. Meaning has been extinguished in what Baudrillard calls "the code"—a self-replicating system of entangled significances so weighty and yet so nebulous that it absorbs and replaces the world itself. A dire analysis, I know, made all the more so by my attempt to give it brief and pointed expression, and yet I think the underlying intuition—that life in the contemporary phantasmagoria loses its integrity, its orientation, and its grounding—is widespread. The question then is: What is to be done? I will come presently to what I think the CBC specifically can do, but first I'd like to look at the responses proposed by the scholars who've helped me define the problem.

Standing in the Flood

Harold Innis knew perfectly well that media have always shaped the style of human awareness—that there has never been a reality without representation, or representation without a distorting medium through which it had to pass. It was his own studies that had first demonstrated this effect. And yet he was convinced that the speed, portability, and reach of contemporary communications media—even those that existed by the time of his death in 1952—were leading his society into a fatal "present-mindedness," and a consequent "neglect of problems of time."[26] Planning and prediction were overshadowing spontaneity and surprise; formal, centralized systems of thought were eclipsing implicit, embodied local knowledge; rigid concepts were replacing the flexibility of conversational exchange; visualized space was taking the place of remembered time. Innis's response was to issue a fervent "plea for time" and a call for the revival "the oral tradition."[27]

The essence of Innis's idea was balance. He knew, as I've said, that the biases imparted by media are inherent and ineradicable, but he felt that they could be played against one another in such a way that one would compensate for the limitations of the other. A society like his, mesmerized, as he thought, by vision and speed, by fixed texts and overdetermined concepts—its consciousness a collage of ever more minute bits of information—needed the leisurely spaces, extended times, and irreducible ambiguities of oral give-and-take. One of Innis's instances of such a balance between counterpoised media was the period between the eighth and fourth centuries BCE in ancient Athens. He thought that during this time oral and literate modes of thought had checked and counter-balanced each other in a way that was exemplary for contemporary societies. His sense of this period as an inspiring prototype of complementarity between media was shared by his friend and colleague Eric Havelock, who summed up their joint opinion in his book *The Crucifixion of Intellectual Man*: "There was a golden age in Athens when men, as they walked the streets lived in two minds at once, guided by the unconscious heroism of an epic tradition, yet roused to vivid thought by the science of an awakening intellect."[28] Innis developed the same idea in the contrast he drew between Plato and Plato's student Aristotle. Plato, Innis said, preserved crucial vestiges of the oral tradition in his "immortal inconclusiveness" and kept the play of ideas-in-conversation alive in his characteristic dialogue form. Aristotle exemplified the growing predominance of writing. He rejected "the mixture of the oral and written traditions" in his master's work and instead "affirmed the absolute monarchy of the mind."[29] Encyclopedic and explicit system began to replace a more flexible, provisional, and dialectical style of philosophy. This reflected what Innis called "a monopoly of knowledge"—a situation of uncompensated bias in which a single mode of communication, along with the class which operates it, achieves an unhealthy preponderance.

Innis felt that crucial cultural capacities were disappearing from his society. Rule-following was replacing judgment; fixed positions were undermining conversation; and urgencies of every kind were reducing the space for unhurried thought. In his view, these one-sided habits of mind were all direct or indirect effects of media—a result of allowing our predominant media to impose an exclusive and, in his terms, monopolistic mode of thought. His prescription: Learn to think in more than one way. Be "in two minds," as his friend Havelock had said, or perhaps in even more than two. Hegemonic media must be counteracted by other media. Centralized monopolies must be challenged from their margins where their mimetic power is weakest.

Marshall McLuhan, Innis's inheritor and acolyte, focused on the careful study of media—the second approach I want to highlight. He adopted as his exemplar and alter ego the figure of the shipwrecked sailor in Edgar Allen Poe's short story, *A Descent into the Maelstrom.*[30] In this story, the sailor tells his rescuers how he survived the whirlpool that had wrecked his ship by observing the properties of the maelstrom and then lashing himself to a cylindrical cask on which he was able to ride it out. Without a knowledge of the structure and syntax of media, akin to the old mariner's feel for the operation of the whirlpool, McLuhan thought, his contemporaries would be little more than animated expressions of their electrified environments. Sometimes his stance confused people, particularly during what American media critic Mark Crispin Miller calls McLuhan's "mock-ecstatic" phase, when McLuhan was often thought to be celebrating the vortex into which he had thrown himself rather than just trying to survive it.[31] But, in his more considered works, the keynote was always study. What was needed, he said, was a "new science" able to revive what he called "the ancient skill of reading."[32]

In speaking of an *ancient* skill, he was referring to an idea that traces back to Plato: that the very foundations of education lie in the study of grammar, rhetoric and dialectic, the three elements of what was called the *trivium* in the medieval university. In the old university curriculum these subjects were considered the very foundation of the liberal arts. Before proceeding to the various fields of knowledge, one first had to understand how knowledge is made. Grammar studied its linguistic infrastructure, rhetoric its form of expression, and dialectic its logic or meaning. Today, all three of these studies are considered inconsequential academic luxuries—grammar refers only to a pointless knowledge of the parts of speech; rhetoric points to talk that is ornamental and insincere; dialectic is the arcane province of professional philosophers.

Abandonment of the *trivium* reflected a foundational modern confidence—the confidence of a modern, scientific society that it had tamed the power of language and learned to give texts a plain and unambiguous meaning. By the time of the Reformation, Martin Luther was already proclaiming that "tradition and interpretation" were just impediments placed between ordinary readers and the Word of God by self-interested authorities.[33] The proto-scientists of the seventeenth century—they called themselves natural philosophers—undertook to purge language of all rhetoric and deceptive artifice and to adopt an exacting and explicit style that would make their meanings plain and unmistakable. Thus was born the *objectivity* whose ghost still haunts us. But McLuhan believed the age of unambiguous meaning had drowned in the "perpetual flux" of the new information environment, where significance "mutates at electric speed." His proposal was to return to the study of grammar and rhetoric, and to expand these studies to take in the entire "information environment" and the ways in which we perceive it. Else, he warned, we really will be living in an autonomous and self-replicating "media ecosystem," as far from our control as "the wind and the tides."

Both McLuhan's new science, and Innis's plea for time, offer promising ways of withstanding the disorientation and light-headedness produced by our current media glut. But I think they need to be supplemented by a third proposal—from fellow philosopher of technology, Ivan Illich. What he adds to Innis and McLuhan's recommendation is the idea of self-limitation, the idea that sometimes it's necessary to just say no.

Like his two predecessors, Illich saw contemporary society as *technogenic*—his word for a world in which the ensemble of tools and techniques that we call technology largely generates the social order. In such a society, he thought, a person's very self-image can be insinuated, often subliminally, by the surrounding techno-sphere. We think of ourselves on the terms proposed by our technical systems—whether they are jet-liners or drones, organ transplants or AI, death cocktails or in-vitro fertilization. We speak of ourselves in metaphors drawn from our cars and our computers, our movies, and our medical instruments. We are exposed, via media, to an unending flood of apocalyptic images—to talk of nuclear war and genetic editing, geo-engineering, and "ethnic cleansing"—constructs of what Illich called an "epistemologically explosive nature" because even to speak of them is already to accept the unacceptable. The result, according to him, was that contemporary people stand in danger of "burn[ing] out [their] hearts."[34]

In Greek mythology, those who dared look into the faces of the dreadful Gorgons were turned to stone. Contemporary persons, Illich thought, have only to turn on their televisions, or open up their "news feed," to be faced with comparably terrifying apparitions, but most of us believe ourselves immune to their power to petrify us. Illich was not so sure. He feared that those who engaged too casually or too continually with today's neo-Gorgons would lose crucial sensitivities and hesitations. He feared that they would become inured to the unspeakable, accustomed to the unthinkable—their sensibilities numbed and calloused. His point can be extended, I think, to our exposure to more everyday sorts of information as well. Whatever overwhelms our capacity for assimilation is apt to produce cynicism, desensitization, and habituation. There can be too much, even of a good thing, as an old proverb says. But, at present, virtues like reticence and restraint, modesty and forbearance, all stand under suspicion of abetting repression. Wherever the searchlight of liberating publicity fails to shine, it is thought, domination will thrive in the darkness. The result is the messianism about news that now pervades the CBC and makes "more news" the almost reflexive answer to every question about how the corporation can better fulfill its mandate.

Illich instead called for a "new asceticism," a new form of the ancient virtue of "guarding the senses." He was not talking of hair shirts or other forms of self-mortification, but of "something profoundly different from any [form of ascetism] previously known."[35] He was searching for a way that he and others could stand in the technogenic flood that he thought was corrupting humanity's image of itself. He could not imagine this stance without a certain renunciation—a refusal, as he says, "to use certain words or to permit certain feelings to creep into my heart." There will be times, he said, when it is better to stand back, or stand aside, rather than to throw oneself into the whirlpool like McLuhan's shipwrecked sailor.

The Way Ahead

I have not meant to imply, by this précis of Innis, McLuhan and Illich's views on how to counteract the contraction of time, the disorientation of perception, and the cauterization of hearts, that these writers have made proposals that could easily be incorporated into a CBC vision statement or action plan. I know that the CBC must find its way, today as throughout its history, in a severely constraining and sometimes actively hostile environment. But I do think that these scholars all point to what must, at least, be a starting point for discussion, and that is *awareness*. Practitioners of media—the CBC among them—have an almost inevitable interest in taking the character of their medium for granted in order to focus on "what needs to be communicated." Forgetfulness, on this score, simplifies their task and underwrites a more heroic and more altruistic accounting of their actions. It is easier to deliver "the news that Canadians need" than to question the procedures by which *the news* becomes news in the first place; easier to boldly present "The World at Six" than to query the hubris that allows a montage of highly stylized snippets of information to be presented as simply *the world*. Stories are more quickly and conveniently created with sturdy clichés, archetypal characters, and time-tested plots than with self-conscious questions, bracketed assumptions, and careful attention to the distorting properties of the medium being used. But our circumstances, as I have been stressing, are exceptional. Immersed in Baudrillard's *simulacrum* and surrounded by figments, fables, and phantoms whose reality we are unequipped to gauge, we are in need, as never before, of a return to earth, sense, and orientation.

This need can only be met by what I would call *an ethic of care*, though I propose the term cautiously and mean by it nothing sentimental. What I'm talking about is a certain solicitude about the consequences of our actions. So long as the CBC forms a cozy "we" with its audiences, and so long as it takes its extraordinary means of access to them for granted, the CBC has no need to care in the sense in which I am speaking. Under its current assumptions, it is only behaving as an innocent component of the "media ecosystem," and only saying what everyone knows. To exercise an ethic of care would be to grow more tentative and more self-aware. Care, as I would like to define it, would be interested but uncertain, attentive but unassuming, apprehensive but unafraid. Care is feeling one's way, mindful of the disturbance one is creating, reluctant to impose. Care is tact.

Political philosopher Leo Strauss was once asked for "a general rule regarding teaching." He replied: "Always assume that there is one silent student in your class who is by far superior to you in head and heart."[36] Such circumspection is another aspect of care. Care is careful. CBC broadcasters ought to assume that they don't know who's listening—no "target audience"—and that the medium they are using comes between them and whomever this might be—no mystifying talk of "your friends at CBC Radio." It's time for the CBC to abandon the myth of transparent communication and stop pretending it's an angelic messenger, bringing news from a world it had no part in creating.

French philosopher Simone Weil offered a parable of mediated communication. "Two prisoners whose cells adjoin," she wrote, "communicate with each other by knocking on the wall. The wall is the thing which separates them, but it is also their means of communication."[37] So it always is: what joins is what separates, what separates is what joins. Recognizing this situation with humility and humour is the first step to a responsible practice of media.

Much has recently been said and written about a supposed crisis of truth. Journalists like Jonathan Rauch have even felt called to mount "a defense of truth."[38] Often discussion turns on whether truth is relative or absolute, constructed or given, as if the choice between these pairs were a decisive shibboleth by which to sort the good people from the bad people. I would prefer to start from the recognition that truth is mediated. This is the case whether or not truth exists somewhere as a pristine idea untouched by grubby human hands. Media-dependence is simply our condition as thinking, speaking beings, and becomes so, *a fortiori*, in the age of technological media. We cannot escape this situation, but we can take better care of it.

9 | After Objectivity: Toward a New Pluralism

We have hitherto considered only two possibilities: that the received opinion may be false and some other opinion, consequently, true; or that, the received opinion being true, a conflict with its opposite error is essential to a clear apprehension and deep feeling of its truth. But there is a commoner case than either of these; when the conflicting doctrines, instead of being one true and the other false, share the truth between them; and the nonconforming opinion is needed to supply the remainder of the truth, of which the received doctrine embodies only a part.

—*John Stuart Mill,* On Liberty[1]

There are no whole truths; all truths are half-truths. It is trying to treat them as whole truths that plays the devil.

—*Alfred North Whitehead*[2]

The truth comes into this world with two faces.

—*Black Elk Speaks*[3]

The idea that I want to discuss in this chapter goes under many names. Objectivity, truth, and information, when opposed to misinformation, are the most important of them. These terms are generally used with the implication that facts are a naturally occurring phenomenon, like trees or stones. It is assumed that such facts are readily distinguished from opinions, and that there is wide agreement on which facts are relevant to a given case. These understandings taken together constitute what might be called the default mode of journalistic self-justification. So, for example, when Christopher Waddell and David Taras in their proposal for a renewed CBC, say that news is "the lifeline of democracy" and must therefore be "the tip of the spear for a new CBC,"[4] they take it for granted that everyone knows what news is. News is comprised of ascertainable facts, recognizable as such by all right-thinking people, and all the CBC needs to renew itself is the resources and the determination to gather and present the facts that Canadians "need to know."

The idea of objectivity, or more grandly, truth, is now widely understood to be in crisis, beset by a flood of misinformation. This crisis became notorious after the election of Donald Trump in 2016, but one can trace the fear of the emergence of a "post-truth" regime back even further—to at least 2004 when a George W. Bush "White House official," widely thought to be his senior adviser Karl Rove, told reporter Ron Susskind about his administration's willingness to leave the "reality-based community" behind in order to "create our own reality."[5] Established media have responded with a campaign to refurbish and reinstate the threatened hegemony of the facts. In 2019, the BBC established the Trusted News Initiative, intended initially to combat "harmful vaccine disinformation," but then expanded in scope.[6] Its "partners" soon comprised a galaxy of information and media companies, including public broadcasters, like the CBC, major press agencies, "newspapers of record," like the *New York Times*, and the tech giants like Google, Meta, Microsoft, etc. "Fact-checking services" multiplied.[7] The *New York Times*, as already mentioned, identified a "reality crisis," and produced a story in which several experts spoke in favour of the appointment of a "reality czar."[8] The United States even announced the creation of a rather terrifying sounding Disinformation Governance Board under the auspices of the Department of Homeland Security.[9]

Journalists rallied to the side of truth. American reporter Jonathan Rauch published *The Constitution of Knowledge: A Defence of Truth*. It described an attack on the "norms" of "liberal science," which had left the institutional architecture of verification on which truth depends beset by "nihilists and bullies."[10] Batya Ungar-Sargon, writing against the deleterious influence of what she calls "woke media," diagnosed a threat to "objectivity" as "one of the central tenets of journalism."[11] Canadian reporter and columnist, Terry Glavin, discerned "a deep epistemic crisis." Our time, he wrote, is marked by "the collapse of consensus about how to go about the work of determining what's true and what isn't." "This is directly related," he continued, "to an increasing tendency across journalism, academia and government policy to conflate knowledge with belief ... a tendency that's fatal to the functioning of liberal democracy."[12] (Glavin made these remarks in the course of a reflection on the abuse he had suffered as a result of a long article he had written—careful, diligent, and respectful, I thought—which tried to sort fact from fiction in the matter of possible grave sites at former Indian Residential Schools in Canada.[13])

I am sympathetic, in various ways, to the view that I think underlies all these statements: the view that something finally is or is not the case. If former residential school sites in Canada contain the unmarked graves of "missing children," as everyone up to the prime minister of Canada has asserted, then sooner or later someone's going to have to produce the remains they supposedly contain. (No one has, so far.[14]) And yet I also think that all the initiatives, and pleas for truth, that I have cited are pointed in the wrong direction: toward an imaginary Good Old Days, when news was trustworthy; there was a consensus about how to tell what's true from what isn't; and "liberal science" consisted of "a community of error-seeking inquirers accountable to each other" with none possessing any "final say."[15]

In saying this, I do not mean to dispute the existence of truth, the importance of evidence, or the virtue of fairness. Certain elementary facts are undeniable, and certain courtesies indispensable. What I want to point out is that most of what appears in a newspaper or is recounted on the CBC is not of this unproblematically factual character. What is more typically being presented is not facts but a *world*, constructed, inevitably, according to the prejudices of the presenters. Such world-making may pass unnoticed in periods of relative social consensus, when the dominant opinion is powerful enough or persuasive enough to induce general agreement and perhaps even produce a rough harmony of interests. But we do not live in such a situation. We live in a situation of extreme dissensus. Canadians disagree on first principles, and they disagree across the board. Whether the subject is abortion or MAID, globalization or degrowth, the plasticity of gender or the purpose of marriage—people are divided on fundamentals and without a common ground on which to negotiate their differences. The more these disagreements are packaged as a matter of sanctioned truth versus certified disinformation, the more aggravated and acute this dissensus is likely to become.

Objectivity is a ruin, the possibility of reinstituting an authoritative regime of truth only a consoling fantasy. What is required of a public broadcaster under these circumstances is not a rearguard defense of a dubious Golden Age but the creation of some quite new: a thorough-going pluralism founded on a commitment to civic peace. By the word *pluralism* I emphatically don't mean relativism, or what philosophers sometimes call value pluralism.[16] I mean, first, respect for views that rest on different grounds than one's own, and, second, a commitment to open discussion and inquiry regarding these grounds. Pluralism, in the sense I would like to use the word, opposes fundamentalism, rather than making a claim about the character of truth. If people disagree, and these disagreements rest on premises which may not even be fully conscious, let alone articulate, then what is needed, it seems to me, is a place to talk, and think, and not the re-imposition of an authoritative doctrine and the ostracization of everyone who won't accept it.

The History of Objectivity

The word *objectivity* has had a long and varied history in the natural sciences. According to historians of science Lorraine Daston and Peter Gallison, these variations have been so great that the term has acquired, over time, a "smear" of significances, rather than one precise denotation.[17] In journalism, according to American media historian Michael Schudson, the word didn't come into widespread use until the twentieth century.[18] Its adoption corresponded with the consolidation and cartelization of the newspaper business in the period around the First World War. In the nineteenth century, says Canadian historian Paul Rutherford, journalism had been a bare-knuckle sport with a wide variety of participants:

> ... there were ... many newspapers appealing to ... many small publics which were supported by various political, and to a lesser extent, clerical and business patrons. Now the thing about newspapers in this period is that it didn't cost much to start them or maintain them. They were really small enterprises. It wasn't big business ... And that meant that all kinds of people could get into the marketplace of ideas. Halifax, for example, in 1864, had eight tri-weekly newspapers. Eight! All reflecting ... different opinions ... [each] committed to their version of truth.[19]

This was the period in which Canada and the United States really had something like the free press, which is supposed to function as a check on the abuse of power in liberal political orders. Corresponding to this regime, Rutherford continues, was a theory of journalism very different than that imagined under the name of objectivity ...

> ... journalists ... did not look upon news as objective fact. There wasn't anything like objective fact. There was truth, and there's a difference between truth and fact. Truth is something that is philosophically defined, and it's defined by your doctrine or ideology. So ... what was fact to a reformer was not necessarily fact to the Tory—never mind to the Irish Catholic, or the French Canadian, or the Presbyterian or the Orangeman, or what have you.[20]

In the early years of the twentieth century, the ideological variety of the nineteenth-century media landscape was replaced by a much more consolidated enterprise. Between 1914 and 1921 alone, forty daily newspapers failed in English Canada, approximately a third of the total number. Some competition continued. *The Star* and *The Telegram* continued to make a great show of their mutual antipathy in the afternoon market in Toronto for another fifty years, until *The Tely*, as it was known, finally failed in 1971. But competition and ideological diversity were much reduced. Driving this consolidation were enlarged markets, increased costs, and the replacement of subscriptions by advertising as the primary way of meeting these costs. Objectivity was the ideology of this new mass-produced media. The collapsing of separate publics into one mass audience required standardization of the product, and the need for universally intelligible stories required the reduction of the social world to easily grasped causes and effects. Anthony Smith (1938–2021), a British broadcaster and media scholar, calls this required simplification "the error of social causality":

> At the root of [the] error of social causality is the concept of the fact, the idea that the world is constituted, can be represented as, can be reconstituted as a collection of facts. Now the notion that the world consists of facts is a very technological notion. Only a highly technology-obsessed, a technology-based type of society could develop the notion, and be at all happy with the notion, that the world consists of facts. When one looks at the history of the debate about objectivity and the history of the processes of news collection, one can see how the concept of fact is a meretricious convenience of an industry that collects and produces news information. Somehow, we have to get away from the idea that there are facts ... that exist beyond the apparatus for reporting them. Beyond the apparatus of reporting there do not exist facts—there exists a world, and the world is not made out of facts.[21]

The issue, of course, is subtle and complicated, as even a quick dip into the history of the fact/value distinction in philosophy will reveal. But I don't think that Anthony Smith is disputing the existence of commonly agreed objects of perception like trees or bicycles, or questioning the procedures by which historians and journalists alike try to discover, as faithfully as they can, what has occurred in a given case. No more is he denying that the atomic number of gold is seventy-nine in Toronto and Tokyo alike, or claiming that all facts are created equal. He is saying only that in a social and political arena like journalism, "facts" will coalesce out of the virtual infinity of worldly goings-on, according to the prejudices, protocols, production requirements, and economic necessities of the media purveying these particular facts. And he is saying that *world* is prior to fact and will inevitably determine which facts are considered salient and relevant.

Mass media, Anthony Smith says, depended on simplified sequences of cause and effect to sell their product. They needed facts out of which to weave these sequences, and they needed objectivity as a legitimating account of their reporters' special ability to recognize which facts were pertinent in a given case. Several generations of media studies have not been kind to this story.[22] Careful attention to what news and media organizations actually do has shown definitively that journalism is as much a creature of its mode of production, and the prejudices of its practitioners, as is every other branch of knowledge.

Journalism is influenced first of all by what comes to its attention, and by the agencies that supply and manipulate this attention. A second form of structural bias arises from the organization of work. News, by definition, involves bits and pieces of information, juxtaposed to one another, with little context. These bits and pieces will only make sense if they involve familiar situations and rudimentary dramatic structures—triumph and tragedy, victory and loss, rise and fall, etc. In the brief, pressurized interval of a newscast, the terms in which events are described must be well understood in advance—there is no time for explanation, and no time for variation within standard terms. Some form of counting will usually provide the main measure of significance. The severity of a disaster will be measured by a death toll. An election will be rendered as a horse race. Politics will reduce to a rudimentary set of conflicts and contrasts. These straight jackets are inherent in the form.

One can only represent the world in the five minutes of *The World This Hour* by constructing a universal scale of significance—a scale by which the producers can weigh the train derailment in Sri Lanka against the riot in Ecuador, or the indisposition of Canada's prime minister against the collapsing price of pork belly futures on the Winnipeg exchange—and such a scale can only be constructed through a whole series of tendentious judgments whose criteria must be concealed in order to preserve the fiction that this is the news. By its nature, the news cannot help being, on the one hand, a disorienting and discontinuous phantasmagoria, and, on the other, a rigidly conventionalized expression of the existing order. These constraints impart a strange irony to the name news because they work to prevent anything genuinely new from ever appearing. The news is instantly rendered old and manageable by the magnetic and inertial power of formatting. The production process itself excludes alternative framings as a matter of necessity.

Back in its heyday, *Mad Magazine* used to parody the *New York Times* prominent boast that it contains "All the News That's Fit to Print" with the claim that *Mad* featured "All the News That Fits the Print." This pretty well sums up the way in which the infrastructure of the news business determines the eventual product. But information media are also influenced by considerations that go far beyond their own practical necessities. They also reflect, and at the same time generate, the larger frameworks within which worlds assemble and begin to make sense.

I first began to notice how these more comprehensive framings take shape, and how they change, when I was the executive producer of the CBC's morning show in British Columbia. The year was 1975, and the issue of the day was the approaching end of nearly thirty years of war in Vietnam.[23] I had come of age politically during the popular mobilization against the American war in Vietnam, and I had long assumed that the war must end in the reunification of Vietnam. But, just as the American puppet regime in South Vietnam finally crumbled, the issue was abruptly reframed. The Vancouver newspapers blazed with banner headlines about "the fall of Saigon." CBC reporter Mike Duffy filed breathless reports from the deck of an American aircraft carrier lying off the coast of South Vietnam. The injustice and unpopularity of the American invasion of Vietnam were set aside. All of a sudden, it seemed that "we" were evacuating, that "we" were the losers in a war which perhaps we ought to have won.[24] I had the impression that I was witnessing a historical "scene change" in which one could practically see the stage revolving, the furniture being rearranged, and the actors changing their places in the darkness.

I have since witnessed several more of these, to me, astonishing transformations. The most recent was the COVID-19 pandemic. What took place around the time a state of pandemic was declared by the WHO was a demonstrably radical change—one might even say a revolution—in the conventional wisdom of public health. But this change went unremarked on the CBC and in other major media. When eminent public health veterans tried to point this out, they were either banned, like Dr. Richard Schabas, or just ignored, like the equally distinguished signers of the open letter to the government calling for a more "balanced approach" in the summer of 2020.[25] Again, it seemed as if a new kind of common sense was suddenly in force and no one could remember that there had ever been any other way of looking at things.

The events of 9/11 were another case of this instantaneous unanimity. The dust had barely settled before everyone seemed to agree that "the world had changed forever" and that a catastrophic War on Terror was an almost inevitable entailment of this change. Going farther back, I think of the strangely oblivious mood in which Europe initiated the Great War in the summer of 1914. "The protagonists in 1914 were sleepwalkers," writes historian Christopher Clark, "watchful but unseeing, haunted by dreams yet blind to the reality of the horror they were about to bring into the world."[26] In all these cases, people inaugurated a new epoch, and a new worldview, as if in a trance, and as if there were no choice whatever in the matter.

Media do not, by themselves, produce the frameworks of which I've been speaking. COVID-19 could not have been anointed the enemy in an all-out war, or served as the occasion for a revolution in public health, had it not incarnated apocalyptic anxieties that had been building up for a generation. You can only tickle the public in the places where it's ticklish. But, with this proviso, it can be said that media enthusiastically reproduce each new consensus—and often with such amazing speed, confidence, and authority that it can sometimes be difficult to tell whether one is looking at the chicken or the egg. In the case of COVID-19, the newspapers I read reinforced the impression of total emergency and total unanimity by carrying almost no other news for months. No one who was being informed by such sources alone could have doubted what one of these newspapers, *The Globe and Mail*, made explicit in September of 2020: that "Canada [was] at war."[27]

Manufacturing Consent

How then are these *new normals*—a phrase frequently heard during the pandemic—from time to time produced? In the case of the 1975 "scene change," I was puzzled for years about what had occurred. The writers who offered me the beginnings of a plausible answer were Noam Chomsky and Edward Herman, in a two-volume work, published in 1979, called *The Political Economy of Human Rights*.[28] What I had been witnessing, these writers claimed was the beginnings of what they called "the reconstruction of imperial ideology." Their argument, in very broad strokes, was as follows: the growth of popular social movements from the 1960s on, and particularly the anti-war mobilization, had created a fear among American elites that democracy was becoming "ungovernable." The problem was identified in a report to the influential Trilateral Commission by political scientists Samuel P. Huntingdon, Michael Crozier, and Joji Watanuki in 1975. Later published as a book, the report was called *The Crisis of Democracy: On the Governability of Democracies*.[29] It argued that too many people had stopped believing in the beneficence and the necessity of American political and military hegemony, and that control over public opinion must be reasserted. Chomsky and Herman describe how this was done. The essential mechanism, they say, was, and is, the application of an unbridgeable distinction between friends and enemies.

Through an analysis of hundreds of newspaper and television reports, Chomsky and Herman established that the misdeeds of foreign governments fall into three broad categories: They call them benign, constructive, and nefarious terror. *Benign* names terror to which the West, broadly speaking, is indifferent. A dramatic example which has occurred since Chomsky and Herman wrote is the war which began in the Democratic Republic of Congo following the 1994 Rwandan genocide and continued on and off for the following fifteen years. A bewildering array of domestic armies, foreign armies, militias, and private security forces killed millions of people—reputable estimates vary from four to six million dead, with perhaps 2,000,000 more displaced, and widespread rape and sexual violence. On its face, this event might loom as large as the Holocaust in the political and moral imagination of the West. Instead, it is little noticed, and, when it is, no wider significance is ascribed to it. No vital interest is seen to be at stake. It is, in Chomsky and Herman's bitterly ironic term, benign terror.

Constructive is their word for terror conducted by friendly and useful client states. Latin America has been scourged by this type of terror throughout my life, as governments trained, backed, and sometimes directly installed by the United States have tortured, "disappeared," and, now and then, massacred their citizens. It is not impossible to find out about such things. A certain amount of hand-wringing, and even distaste for the dictators who underpin the American imperium, may even be allowed to form a sub-theme, or counter-melody, in media coverage. But, the most important thing, finally, will be that these governments are on our side, and their enemies are our enemies. Crimes committed by such regimes will be regrettable, but not particularly salient.

Lastly, there is *nefarious* terror which is committed by enemies and deserves unqualified condemnation. At the moment at which I am writing Russia's invasion of Ukraine provides the best example. Again, just as constructive terror may inspire regrets, doubts, and perhaps even a "hard question" or two, so nefarious terror will not be entirely without extenuation. Those who study events carefully will know that highly placed Americans have been warning since 1990 that military encirclement of Russia would be bound, sooner or later, to provoke a response. Some may even know that Vladimir Putin once petitioned to join NATO. But the overwhelming impression will be of innocence overwhelmed by unaccountable evil.

The subject is difficult to write about, and easily misunderstood. Russia's invasion of Ukraine is evil, and the death and destruction it has caused, as with all war, is hard to contemplate or to bear. But Russia's actions are not uniquely evil, nor any more evil than things we shrug off or barely notice. "Human kind cannot bear very much reality," as T. S. Eliot wrote, and only through a series of reductions can a coherent world be made out of the inexhaustible welter of events.[30] Herman and Chomsky's hypothesis is that the friend/enemy distinction provides the reducing valve by which the news is sorted.

The categories that Herman and Chomsky created nearly fifty years ago still fit the world of today. "Nefarious terror" is evident in Ukraine, where nothing less than "Western civilization" itself is seen to be at stake after an invasion by a recognized enemy. "Constructive terror," is in progress in Gaza and Lebanon, where, despite massive unpopularity, and even some elite misgivings, an ally is still seen to be at work. "Benign terror" can be seen, if one cares to look, in suffering Sudan, where the starvation and displacement of millions of people by a vicious civil war has so far attracted little attention. What matters about this, for my argument in this book, is not how the world, at a given moment is constructed, but just that it is, and must be, constructed. What people see is determined by these framings. Jews who see Israel as the just reclamation of an ancient homeland, and an indispensable last refuge against perennially resurgent anti-Semitism, are looking at the same reality as the Palestinians who see themselves as having been unjustly uprooted and exiled by an expansive settler colony. They don't, however, see the same thing. Such framings can, of course, be challenged, adjusted, and improved. But that does not mean that there is some pure objective factuality waiting to be discovered once the relevant framings have been exposed. The new story, however much improved, will still be partial, self-interested, and itself in need of criticism. Recognizing this partiality, in my view, is not relativism but simply realism about the way in which our assumptions colour, often unconsciously, our understanding.

Recognition of our dependence in this regard, as well as our lack of transparency to ourselves, seems to me to point decisively in the direction of pluralism. There is no longer a definitive story, a decisive objectivity, or a unanimous public. Discordant narrative frameworks must now be encouraged to face one another, interrogate one another, and accommodate one another. Unhappily, the CBC, along with other major media, currently seems pointed in the opposite direction. Mexican poet and essayist Octavio Paz, whom I quoted earlier, says that "when a society finds itself in crisis, it instinctively turns its eyes towards its origin and looks there for a sign."[31] This describes, quite precisely, what appears to me to be going on at the moment. Instead of looking ahead to the world beyond objectivity, the CBC is looking backward and seeking inspiration in journalism's founding myths. Modernity's founding move—its vaunted Scientific Revolution—was the institution of a clear-cut distinction between facts and opinions, or facts and values as would be said later. Knowledge would be stabilized and secured; "the idols of tribe" would be put on the shelf.[32] This distinction has foundered with the discovery of what Harold Innis called *bias*—the inherent and ineradicable spin that is imparted to our thought by our media, by our starting assumptions, by our languages, and by everything else we must inevitably take for granted. A revolution is implied by this discovery, but this revolution threatens the very idea of the news as a simple, single, self-evident thing, accessible to proper journalistic technique and easily tailored to available journalistic formats. How much easier to turn back.

Misinformation

One of the clearest instances of the reactionary mood I've been speaking about is the current campaign against misinformation. When this word is used, as it now is, as a term of art distinguishing certified knowledge from erroneous opinion and ideological fancy, the whole outworn structure of objectivity that I have been analyzing is restored and reinforced. I'm not talking here about whether there are people who believe that Bill Gates has two heads or that the boffins of the World Economic Forum engage in secret satanic rituals. Quite evidently there are such people. I'm talking about the treatment of such fantastical information as a syndrome that can be countered by a regime of sanctioned information or "trusted news" that has been approved by fact-checkers and endorsed by the Disinformation Governance Board.

This new understanding of misinformation established itself during the COVID-19 pandemic, when considerable effort went into identifying, vilifying, and even outlawing this sinister new thing. The CBC, as I showed earlier, allowed no deviation whatsoever from the official narrative. Colleges of medicine threatened and sometimes carried out drastic discipline against doctors who raised questions about the utility of lockdowns or the safety of COVID-19 vaccines. What epidemiologist John Ioannidis called "a dirty war" was launched against scientific dissenters. "Brilliant scientists" were "threatened, abused, and bullied," Ioannidis says, "reputations were systematically devastated and destroyed."[33]

A war against misinformation of this kind might be thought to imply the existence of a stable body of certified information against which deviation is measurable. But this was not actually the case. The official information changed continuously during the pandemic. The idea that the virus might have leaked from a laboratory in Wuhan, for example, was at first proscribed as a conspiracy theory. Peter Daszak, a scientist with direct ties to the Wuhan lab, sent the British medical journal *The Lancet* a statement signed by twenty-seven scientists from nine countries, some of them also interested parties, which "strongly condemn[ed] conspiracy theories suggesting that COVID-19 does not have a natural origin."[34] Science had spoken. Then, a year later, former *New York Times* science reporter Nicholas Wade published a carefully researched article in the *Bulletin of the Atomic Scientists* showing, on a "balance of probabilities," that a lab leak was COVID-19's likeliest origin.[35] The lab leak hypothesis suddenly became arguable, while still retaining, for most, its air of conspiracy.

The official story on vaccines underwent similar twists and reversals. At first, COVID-19 vaccines were said to prevent transmission of the virus, and policies of compulsory vaccination were based on this premise. Then it became clear that the vaccines did not prevent transmission. The policies that had been based on this erroneous premise were retained nonetheless. This pattern of revision in what constituted information, and, therefore, *mis*information repeated itself. By the end, it was an open question which was worse: the popular misinformation circulating on social media, or the official misinformation coming from politicians, established media, and public health bureaucrats.

The term *misinformation*, in my view, is usually a red herring—a figure that distracts attention from the real problem. The important difference between the misinformation circulating amongst the *hoi polloi* and the noble lies put about by the public health authorities is not a matter of factuality but of power and influence. Yes, information is, ultimately, true or false, but it is more commonly encountered as a blend of the two, and a blend, often enough, so ingeniously concocted that the truths rest on the falsehoods and the falsehoods depend on the truths. Under these circumstances, a Ministry of Truth does not seem a promising solution. What seems to me better is to recognize that information is often a secondary matter, when compared with the orientation that lends the information purpose and plausibility. The establishment easily excuses itself for a few scientific stretchers about the value of masks, or the efficacy of vaccines in preventing disease transmission, because it is wholly convinced of the righteousness of its cause. Its lies are so noble as to appear, in its own eyes, as no lies at all. The stories told by anti-establishment forces are oriented in the same way. The Freedom Convoy's abusive animus against Justin Trudeau made the wildest tales about the prime minister's fealty to the World Economic Forum seem as believable as they were satisfying.

A close companion, and sometime synonym, of misinformation, is "conspiracy theory." Like misinformation, it has a variable application. The pandemic, again, affords many telling examples. One such is the memoir published in 2021 by British scientist Jeremy Farrar, then the director of the Wellcome Trust, a powerful British foundation that funds medical research, and currently chief scientist at the World Health Organization (WHO).[36] In his book, Farrar describes the two months in 2020 that preceded the declaration by the WHO in early March that a pandemic was in progress. "During that period," he writes, "I would do things I had never done before: acquire a burner phone, hold clandestine meetings, keep difficult secrets."[37] The clandestine discussions involved Antony Fauci and Francis Collins of the National Institutes of Health (NIH) in the United States, along with other prominent public health officials. Farrar reports that, at this time, he "saw email chatter from scientists in the US suggesting [that] the virus looked almost engineered to infect human cells." He heard from "credible scientists [who were] proposing an incredible, and terrifying, possibility of either an accidental leak from a laboratory or a deliberate release." He and his interlocutors contemplated the "huge coincidence" that the virus had "crop[ped] up in Wuhan, a city with a superlab." They asked themselves, "Could the novel coronavirus be anything to do with 'gain of function' (GOF) studies?" ("Gain of function studies," which consist of deliberately making viruses more effective—for research purposes—had been conducted at the Wuhan lab with the support of the NIH.) They talked about what the public should be told and when. None of this was disclosed until long after the official narrative of the pandemic had acquired unstoppable momentum. But these men functioned, in the plain meaning of the word, as a conspiracy, and Farrar, with his burner phones and "difficult secrets," is perfectly frank about it.

A second example: prominent American vaccinologist Paul Offit has described a meeting in 2021 held to decide whether naturally acquired immunity should be recognized as equivalent to vaccination in the United States.[38] Four other people besides Offit were involved: Anthony Fauci and Frances Collins whom I just introduced, along with Vivek Murthy, the surgeon general, also mentioned earlier, and Rochelle Walensky, the head of the Centers for Disease Control (CDC). Offit argued that, since the very principle of vaccination is immunity through prophylactic infection, it would be wrong, on scientific grounds, not to recognize naturally acquired immunity. The others held that the promotion of vaccination might be hindered if natural immunity were to be accorded any validity. Offit was outvoted, and this group advised President Biden that only vaccination should be recognized as conferring protection against getting and spreading SARS-CoV-2. On this basis the American government created mandates that cost people their jobs and important civil rights. If I hadn't heard Offit himself tell this story, I would find it hard to believe. "Top American public health officials advise president to ignore science in order to promote an under-studied vaccine that has given no evidence of limiting the spread of the disease"—yeah, sure. And, even if the decision were justifiable, as some might think, it would still be a conspiracy—what else would you call five experts deciding behind closed doors what will be called "science," and what will dramatically change the lives of millions of their fellow citizens?

The point, of course, is that there are conspiracies and conspiracies—some demonstrably real, some fanciful. Now and then, one that has been condemned as fanciful changes status. This happened with the lab leak theory of SARS-CoV-2's origins, as I related earlier. But, by the time the story changed, the official narrative was so well entrenched that it made little difference that science might have invented the peril it was so heroically combatting. People believe what they find plausible, and plausibility is conferred by repetition. The difference between a conspiracy and a conspiracy theory is often a matter of timing, perspective, or power. But the idea has a remarkable disciplinary power. People make up stories according to their degree of sophistication and their degree of alienation, and those who are most alienated and least sophisticated will inevitably make up some whoppers. These wild tales can then be used to overshadow, undermine, and discredit news of real conspiracies. The idea that COVID-19 was engineered in a laboratory can be made to appear no more believable than the claim that the moon landing was staged on a backlot in Hollywood. Confusion reigns.

Current attempts to outlaw misinformation and reinstate trusted news are also related to the current anxiety about the transformational power of the internet. Discussion of this power can be broken down into two phases. In the first phase, a mood of profound, sometimes utopian hopefulness prevailed. The internet was seen as providing, at last, the infrastructure of democracy. People could connect with one another without needing to rely on interested intermediaries. All the objections that had been put against the idea of *the public* from Kierkegaard on—that it was an empty, abstract, and mechanical fiction—could now be addressed. The public finally had in its own hands the means of being a public and need no longer be the "phantom" of which Walter Lippman had written—the public that was endlessly invoked and deployed without ever quite being there in its own right. A new kind of political hope was expressed. From Occupy to the contemporaneous Arab Spring, people began to assemble in their common capacity as people rather than as the vanguards of some political ideology—again as if the public had finally come into its own and the future was today, not tomorrow.

The second period of internet discussion has been as dire as the first was sanguine. Typical of this second period is Soshanna Zuboff's book *The Age of Surveillance Capitalism* published in 2019. Zuboff identifies "a new species of power," enabled by digital media, which consists in the "instrumentation and instrumentalization of behaviour for the purpose of modification, prediction, monetization and control."[39] The new regime, she says, "renders experience into data," by "extracting" valuable information from "your bloodstream and your bed, your breakfast conversation, your commute, your run, your refrigerator, your parking space, your living room."[40] Zuboff's book is overlong and sometimes overwrought, exaggerating the impression of novelty it wants to convey by giving old forms of exploitation new names, but, in the end, she is persuasive that digital media have opened capitalism's final frontier by "plumbing the intimate patterns of the self."[41] Her book is just one of many that has explored the dystopian potentials of the internet in recent years. The misinformation panic constitutes another strand in this discourse.

The internet has enabled people to talk among themselves as never before. So many people are now producing podcasts that there may soon be no one left to listen. With such an increase in the volume of communication, folly is bound to expand with wisdom, and lies with truth. Indeed folly and mistruth are likely to have the edge, just because their stories are simpler and more streamlined. Even so, it's worth asking a question posed by American journalist Joseph Bernstein, in a recent article in *Harper's*: "Is social media," he asks, "creating new types of people, or simply revealing long-obscured types of people to a segment of the public unaccustomed to seeing them?"[42] My answer would be that both his alternatives are currently in evidence. Media do have a formative influence on how people reveal themselves, and in that sense, the internet is creating "new types of people." But I think it's also the case that a lot of formerly voiceless people have been given a voice by social media. Given the tools to think and speak for themselves, people are doing so. However defective the results may sometimes be, the right response is not to try to shut them up again, as our prime minister did when he denounced the members of the Freedom Convoy as a "a fringe minority … holding unacceptable views." Slander and sedition should certainly remain illegal, but where stories have been invented, or even just stretched, to give expression to some resentment, suspicion, or injury, it seems better to engage with the cause than to outlaw its symptom—to start a conversation rather than trying to outlaw the public expression of unwanted opinions.

As things currently stand, the misinformation police criticize only unacceptable myths and seem perfectly innocent of the idea that they might have a mythology of their own. There is certainly a lot to criticize, but it seems to me better, in general, to try and understand why stories get made up rather than trying to stamp them out. Take the World Economic Forum (WEF), an organization that seems to inspire the kind of stories that get classified as misinformation. I have heard the president and founder of this organization, Klaus Schwab, boast that Canadian Prime Minister Justin Trudeau and more than half his cabinet have undergone the training that the WEF provides through its young leaders program. The WEF, he said, had "penetrated" Canada's cabinet.[43] Does this mean the WEF is running Canada? No. Does it mean that Trudeau and his ministers are steeped in the globalist, Great Reset, Build Back Better orientation that the WEF fosters? Yes. If some innocent, building on this orientation, then claims that Klaus Schwab has a back channel to the prime minister through which he tells him what to do, then, to me, what has happened is that something that is felt, with some justice, to be true, has been condensed into a handy, easily shared fable. This is not a harmless process, but it is common, and sufficiently inevitable that attempting to outlaw it will create harms vastly greater than the confabulation itself. Dialogue with the teller of such a fable should begin with what he or she is trying to say, not with the exaggerations and inaccuracies that may have marred the telling.

The Power of Myth

What has to be recognized, as I have been emphasizing throughout, is that our present society is divided on first principles. Fellow citizens live side by side in different conceptions of time, in different images of political order, and in different structures of authority. These are differences on the level of cosmology, or mythology, not of information. By mythology I don't mean what is untrue, but what is *prior* to any idea, ideology, or articulation of thought. Myth, in this way of speaking, is what shapes raw experience into an intelligible story. It is the horizon within which things come to light. It is what is looking rather than what is seen—the container not the content—the ground against which figures stand out. If we take the example of science—and *science* is a big bone of contention in the misinformation debate, since it is supposed to determine truth—I would say the *myth* of science is that the world is truly known by taking it apart and reducing it to its underlying components. This is a foundational commitment, a postulate that stands before any observation, experiment, or theory. The *myth* of objectivity has the same character. It supposes as a founding (and, therefore, unquestionable) axiom that *reality* is what is discovered when the world is divided into subjects and objects and the *subjectivity* produced by this cut is then bracketed.

Speaking in this sense, we can say that all knowledge is given shape, impetus, and orientation by myth. This matters to the misinformation debate in two ways. The first arises from the fact that other people's myths are always a lot easier to see than one's own—an effect that is enhanced when those other people are less sophisticated than the ones pointing the finger, and less skillful at disguising their mythologies. It follows that if there is to be fruitful discussion, all myths have to be put on the table, and not just the ones that the mighty have classified as generators of misinformation. The second important point is that relative civil peace will require that antagonists "compare mythologies" rather than abusing each other with facts that have little meaning when separated from their sustaining stories.

Thought is inevitably shaped by certain predispositions. But I don't think that this recognition has to be taken as derogating in any way from the value of science, or as denying the existence of truth. Science does not become invalid when it is apprehended as a particular way of seeing and knowing the world. All that happens is that certain conditions are attached to its validity. Science loses only the privilege it has arguably exercised for the last 400 years of being able to claim that it sees the world as God sees it.[44] It does not lose its irreplaceable usefulness within its proper domain. The same could be said for journalism's struggle for objectivity, in that word's common sense meaning of bracketing prejudice. Some of journalism's self-interested and self-dramatizing poses may suffer from a fuller accounting of bias of all kinds, but its efforts to describe the subjects it takes up in a fair-minded way are made no less worthwhile.

Presupposition, myth, prejudice—all are facets, and versions, of a primal and inherent incapacity—our inability to see those things which enable us to see in the first place. Systematic, across-the-board recognition of this limitation is, for me, the great discovery of our time, echoing through all branches of knowledge and all divisions of science. Incorporation of this recognition into public broadcasting is the only way open to the CBC, if it wants to help reunite our fatally divided polity and recover a spirit of dialogue and civic concord. Critics, of many ideological stripes, are currently preaching a gospel of reaction—whether they want to make America great *again*, or restore "Enlightenment values."[45] But there is no way back. And this is true whether or not the lost world of "trusted news" and "liberal science" ever glowed quite as brightly as these critics claim.

Recognizing multiplicity needn't entail a claim that all stories are beyond criticism and that each person has their own truth. In his 2003 Massey Lectures—Canadian writer Thomas King asserted that "the truth about stories is that that's *all that we are*."[46] [my italics] If this were so, then it would follow that stories are essentially incomparable—each one generating its own horizon—and, therefore, effectively beyond criticism. But, to me, King's proposition is only half the truth. Assuming that King uses the term *story* in the same sense as I just spoke of myth, I think he is certainly right to recognize myth's originality, irrepressibility, and inexhaustible fecundity. But stories are not all that we are, even if they are an undeniable, irreplaceable, and complementary part of what we are. What is left out is thinking. Thinking is the other side of myth/story—its opposite, and its indispensable complement.

Thinking, among other things, is our capacity to stand back from stories, our ability to put them into critical and comparative perspective. Together, myth and critical thinking form a couplet, an irreducible pair, neither of which should be allowed to dominate the other. They must be kept in balance and in constant tension. We cannot allow that the truth is simply the most powerful or the most persuasive story—but neither can we take refuge in some ideal truth that pretends to have escaped the limitations of narrativity. Thought is always to some degree preformed and predestined, grown from roots it does not know and attracted to destinations it cannot foresee. There is always what American poet Wallace Stevens calls "the myth before the myth began."[47] So thinking must always be tempered by the recognition of its mythic roots, and myth must, in turn, be tempered by thinking's constant effort to bring these roots into the light. Neither is all that we are.

Pluralism

Pluralism is the name I have chosen for a style of broadcasting in which contending mythologies are invited to acknowledge and respect one another, and all are subject to analysis and critique. This new philosophy must be based on a firm refusal of the idea that order and authority can be restored by a reversion to faded certainties, like objectivity. Its priority must be conversation; its keystone must be the recognition of difference. By difference I do not mean that luxury of *differences* that currently goes under the name of *diversity* and is often no more than a wildly variegated sameness. I mean difference that grows out of fundamentally opposed worldviews and requires a pluralism that is plural, so to speak, all the way down.

Many modern philosophies have transposed the Christian hope of "a new heaven and a new earth" into a secular key.[48] From Descartes' "masters and possessors of nature" to Engels "wither[ing] away" of the state," a consummation of history has been promised.[49] A perfected and unified society was always just a revolution away; and, so long as this goal was in view, no plan for stable and abiding co-habitation, or even in-habitation, of the earth seemed necessary. Our time has begun, but barely, to think differently. I have already mentioned French philosopher of science Bruno Latour's claim that the modern nations that once conceived themselves as unified bodies have now disintegrated into quite distinct tribes whose cultural and intellectual styles qualify them as separate and often antipathetic peoples. This creates, Latour says, a "diplomatic obligation to introduce ourselves to one another."[50] His intuition that peace depends on the recognition of ineradicable conflict is shared by a number of contemporary political thinkers. To name just two, American William Connolly has written of "agonistic democracy," and French thinker Chantal Mouffe of "agonism."[51] (Both names draw on the ancient Greek word for contest, *agōn*.) These thinkers reject a view of political conflict as always on the way to resolution, and of difference as always on the way to identity. They don't await a day on which all points of view will be brought into alignment. They don't believe, in Mouffe's words, "that it is enough to lift the weight of ideology in order to bring about a new order."[52] Rather they seek "a set of democratic procedures" within which competing views—"hegemonic projects," in Mouffe's terms—can vie with one another without war or social dissolution.

What would happen to the news if the CBC were to abandon the idea of a single, authorized version of things and adopt a pluralist philosophy? The first thing to say is that I don't see why this should threaten any of the genuine virtues currently recognized in the CBC's handbook of journalistic standards and practices.[53] Agreement about what is to be recognized as evidence would remain the foundation of democratic deliberation. Truthfulness and accuracy, impartiality and fair-mindedness would remain pearls beyond price. Evidence would remain the cornerstone of argument. One would simply strip the news of its transcendental claim to objectivity. One would give up the pretense that events present themselves in brightly marked packages with prescribed meanings—the illusion that the news makes and declares itself, and that all the CBC does is to "follow" the events that are "making news," asking the questions that "need to be asked," telling the stories that Canadians "need to hear," and attending to whatever is bubbling up in the "media ecosystem." The founding assumption of pluralism would be that the news is made—by its modes of production, its narrative conventions, and its structuring myths—and, therefore, might be made differently. The constructive, world-making side of newsgathering must be brought out from behind the façade of immaculate conception and made the foundation of a contestable and responsible practice.

This will certainly not spell the end of news. Lots of events, by broad consensus, are news. If the Queen dies, the prime minister visits France, or wildfires encroach on Australia's largest city, the CBC has an obligation as a broadcaster to let people know. I'm saying only that this function should be exercised in as transparent a spirit as possible. News, after all, has been gathered and presented in very different ways since the first printed newsletters began to appear in the sixteenth century, and, sometimes, it has been presented in a critical and self-conscious spirit, with its mode of production and principles of selection made clear.

Pluralism will lay the foundations for conversation. But for conversation to actually take place two conditions must be met. The first is recognition that there is something to talk about—that there is a question to which the parties to the conversation do not think they know the answer before they begin. I believe this to be everyone's condition in the present state of our world. Our civilization, in its blind and unconscious dynamism, has produced a situation with no precedent and, therefore, posed a question to which we cannot yet possibly know some definitive answer. The radically different attitudes toward this singular situation that are now in play call out for conversation. The second condition is that the parties to the conversation must respect its authority as a conversation. Real listening and real surprise will occur only when they recognize that they are in the presence of a possibility that exceeds the conversation's starting points. "To conduct a conversation," says one student of the art, "means to allow oneself to be conducted by the subject matter to which the partners in the dialogue are oriented … Dialectic consists not in trying to discover the weakness of what is said, but in bringing out its real strength."[54]

I have called the new horizon that I imagine for the CBC *pluralism*, but it's not a name to which I'm particularly attached. What matters is this: The CBC must recognize that it has become a monoculture that actively excludes competing points of view and depresses intellectual inquiry. It's time to open the windows, and the doors; time for the national broadcaster to begin to curate a national conversation that is open to the past, open to the future, and open to all who are willing to participate in the requisite spirit of give-and-take. This will not happen until the CBC itself begins to exemplify the moral and intellectual virtues that such a conversation will require. First among these, for me, will be a spirit of adventure—a replacing of the present mood of dogmatism, self-righteousness, and self-satisfaction by a sense of something still to be discovered. I know this prospect seems, at the moment, remote, but, in the long run, peace, amity, and a habitable future will require nothing less. Why not at least begin? "A nine-tiered tower" begins as nothing more than "a handful of earth," says Lao Tzu, and "a thousand-mile journey [is born] … here at your feet."[55] The first task will be to try to think differently.

10 | Living in the Question

> … our own prejudice is properly brought into play by being put at risk. Only by being given full play is it able to experience the other's claim to truth and make it possible for him to have full play himself.
>
> —*Hans Georg Gadamer*[1]

> In actual conversation, something happens. I do not know in advance what the other will say to me because I myself do not even know what I am going to say; perhaps not even whether I'm going to say anything at all.
>
> —*Franz Rosenzweig*[2]

In the medieval legend of the quest for the Holy Grail, the sought-after cup is guarded by a wounded king who rules over a blighted land. He awaits the one who will ask him the right question. Only when this question is asked, will his wound be healed and his land restored. Versions of the story differ on whether the requisite question concerns the king's suffering or the nature of the Grail. A few specify that the questioner must be an innocent who asks from no ulterior motive. But all agree on the necessity of the question. "Deliverance [is] the result of the right question," say Emma Jung and Marie-Louise von Franz in their book on the grail legend, and this "motif," they say, is a universal figure, to be found in fairy tales and legends throughout the world.[3] Canada now waits to be asked the right question, or questions, since the one right question in the grail story is obviously an archetype meant to bespeak a wellspring of many questions rather than a magic formula. The questioner ought to be the CBC, which has been appointed by law for the purpose of "informing" and "enlightening" its country.

When Graham Spry testified in Parliament in 1932, he told the committee charged with determining Canada's broadcasting policy that only a public system would be capable of "cultivating" and "expressing" Canada's "spirit …

character and soul."[4] That Canada possessed these attributes, of soul, character, and spirit, and that a public broadcasting corporation was necessary to express them, was Spry's abiding, though often disappointed, conviction. Few today would dare speak of Canada's soul, or of its destiny, as Spry did on another occasion.[5] The country has grown querulous, fractious, and unsure of itself. A substantial part of the political left doubts the very legitimacy of the Canadian state, calling it a product of genocide. The political right tends to see what Jordan Peterson recently called a "sinking ship."[6] The centre implied by Spry's image of a national soul barely exists. Unity cannot be reimposed—not in the absence of mutual recognition by the warring parties—but it can perhaps be invited. The instrument of this invitation can only be the posing of real questions—by which I mean quite simply: questions to which we do not yet know the answer.

But isn't the CBC already made of questions? CBC Radio mostly consists of interviews, and so does much of television as well. Yes, but these questions very often are not what I would call real questions. Many are so conventionalized that they answer themselves. The shooting scripts for some television documentaries specify the expected answers in advance, with the questions serving only as triggers. Radio interviews are often planned and pre-scripted as a cascade of questions that will be mechanically executed with only minimal regard for the responses they elicit. These are questions of what might be called an instrumental kind—they are means to a predetermined end, accessories to some all-determining format.

I am talking about questions of a more fundamental and more fertile kind. Such questions have two prime characteristics. The first is openness. Open questions do not contain their answers. Rather they open a way into a subject—establishing a direction in which conversation can unfold, and a horizon within which it can take shape. The second attribute of fruitful questions is that they are shaped by the subject of their inquiry as much as by the interests of the inquirer. A true question, says German philosopher Hans Georg Gadamer, "belongs to the being of that which is understood."[7] What he means is that we can't *understand* things if we are not willing to give ourselves to them. Questions must find their footing within the subjects of their inquiry, and not within the dominating and directing will of the inquirer. In conversation, the flow of talk must be oriented by the themes it takes up and not just by the purposes of the participants. Conversations, and subjects of conversations, must acquire independent authority and be allowed to establish their own rhythms, their own directions, and their own impetus toward the openings in which new ideas can potentially flower, and dead-locked points of view seek common ground. "A genuine conversation," Gadamer writes, "is never the one we wanted to 'conduct.' We fall into it—the parties less the leaders than the led. It has a spirit of its own and 'understanding' or its failure is like an event that happens to us."[8]

More of what I mean by questions that show openness and *belong to what is to be understood* can be illustrated by way of contrast with the way things have been done during the populist era at the CBC. During this time, the will of the interviewer/inquirer has generally been paramount. Let me take as an example the account Patrick Watson gives of his style of interviewing and program-making in his autobiography, *This Hour Has Seven Decades*. No better example could be given, since Watson not only helped to define the style that took over the CBC in the 1960s and afterward, he also executed it with extraordinary charm, intelligence, and conviction.

Watson located his authority in the audience—a key tenet of populism. "To be boring or trivial," he writes, "was to be contemptuous of your audience. To make anything public that is of any value … you have to respect your audience."[9] "Holding the audience's attention" was paramount, even if it sometimes required the sacrifice of narrative coherence or "analytical significance." The broadcaster's job, he says, is to compel attention:

> Your task is to create an experience. In a movie or a play or indeed a documentary it is a *world* you're creating. You must make that world perfectly and constantly consistent within its frame. You must do everything you can to make it impossible for your reader or your audience to withdraw, to step out of that world, until the work is complete.[10]

This ambition to "create an experience," in service to the audience, extends to Watson's philosophy of interviewing, which he expounds as follows:

> … more than anything else you must listen. Listen to and intently watch your subject. Follow the leads offered to you. Be prepared to abandon your plan whenever you detect an unexpected opportunity for a spontaneous and thus likely honest exchange or response. And the second great lesson in a sense subtends the first. It is that the subject or subjects being discussed in the interview, what I call "the referential content" is of only secondary interest. The prime interest is the person being interviewed and the dynamics of the relationship with the interviewer.[11]

"Listening" here is a technique. It is not an attempt to discern and follow what Gadamer calls the spirit of a conversation, but rather a quest for the telling moment which will reveal the interviewee to "the audience." (Readers may recall Watson's "periscope camera" with its "low angle extreme close-ups" of discomfited guests at CBOT in Ottawa in 1960, or "the hot seat" on which *Seven Days* staged its interviews with those "prepared to be grilled.") The subject—the "referential content"—is given no independent authority, nor is the interviewee's intention. The "leads" that are offered are treated only as an opportunity for the interviewer to heighten the dramatic intensity of the encounter. The occasion must be shaped, above all, by the question that Watson says he first heard from his "mentor" at CBC Television, producer Ross McLean: "How will it serve the audience?"

What is notable here is that the audience is always invoked en masse and always has a dual character. It is, on the one hand, a kind of animal spirit responding reflexively to dramatic cues—its attention ever apt to wander should the flow of absorbing images be staunched—and, on the other hand, it is the respected public, the people, the highest authority which must at all costs be served. The tastes, wishes, and needs of this creature—primitive and elevated by turns—produce a discipline and morality within which the broadcaster works.[12] The public broadcaster arouses active citizenship by first treating the audience as a mass of easily distracted consumers. This ambiguity is never really addressed in Watson's book, and the reason it doesn't come up, I think, is that the broadcaster's goodwill and good intentions are so much taken for granted. Once the audience's attention is gained, it will of course be turned to worthy objects by the one "creating the experience" on the audience's behalf. Long before Watson, and his colleague Douglas Leiterman, created *This Hour Has Seven Days*, they were already reflecting on "the power of television" and how it "could be used," in Leiterman's words, "to engineer social change."[13] That this power would submit to their good intentions was taken for granted, as was the idea that something called society is capable of being "engineered" in the first place. This was the feature of populism which had its brilliant first flowering on *Seven Days*—the host as the audience's delegate. Just as the leader incarnates the people in political forms of populism, so the host incarnates the people in its broadcast form. Implied is the ability to discern and execute the people's will even to the point of engineering social change.

The assumptions on which Patrick Watson based his theory of broadcasting—disciplined service to "the audience" for whom one could "create experience"—no longer hold. It is in this new situation that we need the non-instrumental, open-ended style of questioning that I spoke of earlier. Watson never spoke quite so bluntly as did Leiterman of "engineering social change," but his memoir makes clear enough that serving the public meant manipulating the public for their own good. He does not approach broadcasting as an arena in which something not yet known is to be discovered but rather as a way in which predetermined ends can be achieved by the broadcaster's skill in creating compelling experiences. The destination to which the audience is to be guided, the understanding to which it is to be brought, is something the populist broadcaster understands *a priori*. The questions he/she asks are not of the kind I described—a mutual opening addressed to an unforeseeable end. They are rather a preconceived means to bring matters to a premeditated conclusion.

Consider Hans Georg Gadamer's statement at the beginning of this chapter: "… our own prejudice is properly brought into play by being put at risk. Only by being given full play is it able to experience the other's claim to truth and make it possible for him to have full play himself." Patrick Watson did not put his own prejudices at risk in the style of interviewing he described in his autobiography. As a man in service to "the audience," his prejudices were sanctified, and therefore untouchable. They were not, as Gadamer says, in play. Watson looked for "leads" and "opportunities" but presented none. He was, in effect, conducting a prosecution on behalf of the audience. So long as his interlocutor shared the prejudices he took for granted and kept out of "play," the conversation might give every appearance of geniality and expansiveness, but Ross Mclean's crucial question—how does this serve the audience?—was always at hand, ready to turn the easy chair into the hot seat.

Gadamer's claim is that the interlocutor must have "full play," and that this can only happen when both parties to a conversation are willing to expose and investigate their own predispositions and preconceptions. This cannot really take place under populist assumptions. I have taken advantage of the frankness and clarity with which Patrick Watson sets out the populist position in his book, but what I have said applies, by now, to several generations of CBC hosts and producers. So long as the host is considered the champion of the audience, and the audience is put in the position of the customer who is never wrong, there will be no place on the CBC for voices that fall outside the consensus in which host and audience are presumed to share—except, of course, on the hot seat. This consensus has tended in recent years to become steadily more narrow and more defensive. It now excludes many "unacceptable" Canadians and tends to cast them as enemies. What was optimistic and outgoing, in the hands of people like Patrick Watson, has become cramped and confining. Prejudice, rather than being put into play, has instead tightened its grip.

Why the Right Question Is Needed

I have argued throughout this book that Canada faces a growing mismatch between the backward-looking but still hallowed discourses that dominate its public discussion and the unprecedented, still barely imaginable situation in which it actually lives. This discrepancy is a prime reason why the CBC must now allow itself to be guided by questions without predetermined answers. Questions, as I've said, can lead onto open ground where new perspectives and new solidarities may show up. They can also help to break apart the hardened abstractions that currently clog our political and intellectual milieu. Much public talk, at the CBC as elsewhere, turns on the existence of problems and solutions, with the problems often described as crises, as well. These problems are not ordinary troubles or concrete difficulties that need to be faced and overcome or avoided. They are gigantic agglomerations that function in public talk as something like staple commodities—the vicissitudes of the economy, let's say, or the problem of homelessness, reconciliation, or climate change. These are, first of all, stupefying abstractions, gathering together and consolidating an unthinkable variety of disparate stories and circumstances. They are, beyond that, resources—the property of the agencies that, in effect, extract the resource and depend on the periodic discovery of new deposits to stay in business. And, finally, problems are syndromes that are defined, and confined, by the available solutions. Just as illness, in medicine, is often constructed to specifications derived from the available cures, so problems in journalistic discourse take shape in relation to preformed solutions—the economy is defined by its imagined policy levers, the drug emergency is constructed in terms of the available treatment options, the health care crisis is seen to vary with the amount of money poured into it, and so on. In all cases, a mould or pattern is created—which then begins to overshadow and retrain individual instances of the supposed problem. One can't avoid naming things, of course, or recognizing similarities between them, but I'm talking here about something more extreme: the transformation of predicaments—which individuals or communities might address—into industrial-strength problems which are projected at a scale at which they can scarcely be conceived, let alone remedied. A problem of this kind allows no question. It gathers overhead like a dark cloud, presenting no precise definition or imaginable way out. It can move money, provoke political passion, mobilize opinion, and create jobs, but only in the interest of its continuation as a problem.

A real question—a question possessing sense and direction—presumes that there is something to be understood and something, moreover, that can be understood. Problems inhibit understanding. They are inert and senseless. Talk of the homeless crisis, the climate emergency, or systemic racism invites a response on the same level of generality on which the problem is formulated. "Systems" must be "put in place." Personal actions that are within reach seem paltry by comparison. One becomes, willy-nilly, a citizen of utopia. Questioning, as I have been emphasizing, opens a path. *Problems* are a pathless maze—a game played by those whose livelihood and self-respect are tied up with their continuance. Only by putting the language of problems and solutions into question will a livable world begin to appear and show its unthought possibilities.

Categories that hide what is yet to be understood, by making it appear as already understood, are the first great difficulty that stands in the way of what I've called living in the question. The second great obstacle is the inhibiting and censorious atmosphere in which public discussion is now conducted. Questions suppose, obviously enough, that the matters to which they're directed are questionable, but that is no longer the case in many areas of inquiry. This is true, most crucially, in the case of history. With churches burning, John A. Macdonald's effigies mostly in storage, or boarded up, and Egerton Ryerson beheaded, besmirched, and disowned by the university that once bore his name, Canada's past is currently undergoing an ideological confinement that allows it to speak, when it speaks at all, only in terms of remorse, revision, and restitution. So long as this foreclosure of historical inquiry and interpretation is in force, nothing like the questioning mood that I am urging will be possible. I will show why this is so in a moment, but first let me establish briefly that the study of history has been disabled in the way that I claim.

In June 2023, the Canadian government's special interlocutor on residential school issues, Eleanore Sunchild, called for "residential school denialism" to be added to the Criminal Code. The minister of justice at the time, David Lametti, responded that "he was open to outlawing residential school denialism with criminal and civil measures similar to those used to punish people who deny, minimize or condone the Holocaust."[14] Ms. Sunchild didn't define denialism, but one could fairly infer, I think, that the term indicted, among others, anyone who refused to admit that the intention and practice of residential schooling in Canada had been genocidal, and that disturbed ground in the vicinity of any residential school necessarily indicated an unmarked grave. Her plain intention was to propose and make enforceable a single authorized version of history. Her premise is that people who went to residential schools effectively own the history of those schools, and that any interpretation of that history other than theirs should be prohibited—by law if necessary. Many other contemporary cases have a similar character.

I believe this move to an authorized history, belonging by right to whoever has the best claim to have suffered from it, will be fatal to a questioning spirit, and not because I doubt the reality of the suffering or have hardened my heart against it. All questions concern the past, because the past, in effect, is all that we have. What, after all, is the present but the guise in which the past currently shows itself—the past's leading edge as it becomes the future? Our words and our thoughts, our institutions and our traditions, are all rivers that rise in past and continue in the present. The past, says historian R. G. Collingwood, is "alive and active" at every moment."[15]

An active past—a past which is present—must necessarily be a changeable past. What changes is not what happened, which, as far as it can be ascertained, cannot be altered, but its significance, which will change as the present changes. The past, to put it another way, has a definite shape, but no definite outcome. We have, says American historian of religion James Carse, "no way of knowing what has been begun there"—because that is something that only the present can determine. As the present changes, Carse says, the past will reveal "new beginnings."[16]

The possibility of finding new beginnings in the past is history's saving grace. But, once history is assigned a fixed and unchallengeable meaning, the space in which we might interrogate the past in search of such beginnings is put off limits. The past is turned into a scene of judgment rather than of living inquiry. Understanding is equated with approval—with complicity in a traumatic story—and, on that basis, outlawed. In the end, there is nothing left *to* understand, just something to be set aside, torn down, renamed, or otherwise got over in the interest of healing. An intelligible past is replaced by a caricature, sketched with a few bold and telling strokes. People who thought and acted as we do, blindly feeling their way across the "darkling plain" of their time, are replaced by a morality play in which competing idealizations do battle.[17]

The future begins in the past, and an open future can only begin in a still undetermined past—a past in which new eyes might discern new beginnings. This matters, at the present moment, because we are the heirs of a tradition that has grown uncertain of its meaning and unsure of itself. I refer broadly to the Western heritage and tradition which, for good and ill, has given Canada its geographic shape, its institutional architecture, and its cultural atmosphere.[18] This tradition has now "lapsed into incoherence," in the words of philosopher Alasdair MacIntyre.[19] What MacIntyre calls *incoherence* is just the situation I have been trying to describe—a situation characterized by basic disagreements about where we stand, where we are going, and who knows what. Tradition, according to MacIntyre, has a narrative structure—it's a story, and, as such, it undergoes what he calls "argumentative retellings." These retellings can run into conflict with one another, and this conflict at times may grow so acute that the tradition, as I quoted, "lapses into incoherence." A tradition thus at cross purposes "can only be recovered," MacIntyre writes, "by a radical reconstitution."[20] This is accomplished not by abandoning the old story but by including it in a new story. "The old narrative," MacIntyre says, "is made the subject of a new narrative."[21]

MacIntyre first made this argument in a discussion of philosopher of science Thomas Kuhn's theory of scientific change, but I think it fits the context I am developing here just as well. MacIntyre's "radical reconstitution" seems to me to correspond quite closely with James Carse's "new beginnings." In both cases, something quite new is being made, but it is being made by reimagining and reworking what has been handed over to us as tradition. Western revolutions have often aspired, catastrophically, to radically new beginnings—the French Revolution restarting the calendar on day one, René Descartes supposing that he could doubt everything, etc. Today's woke utopians are arguably still in the grip of this "paradise-now" transposition of Christian eschatology. I prefer the more modest aim of rethinking the past—because it sticks to the materials we have actually been given.

Such a rethinking is now urgent. Most of the fruitless arguments, and bitter recriminations, that now vex our politics have their roots buried in the past—out of sight and out of mind. But we will be powerless to unearth them without free access to the past, and an open mind about what we might discover there. And this is what is so worrying about the current prevalence of an ideologically contrived history, promulgated by a new class of *commissars* determined to outlaw incorrect versions of the past. There can be no reconstitution of our tradition, so long as our history is allowed to speak only in approved terms. The question we most need to ask people in the past, says R. G. Collingwood, is "What were you thinking?"[22] But we cannot ask this question until we cease making history our scapegoat and begin to understand that we do not yet know "what has been begun there."

Living in the Question

A recent survey of "viewpoint diversity" at Canadian universities found that these once-liberal institutions have turned into "political monoliths." Amongst faculty, self-identified "progressives" outnumber "conservatives" by a margin of approximately nine to one, and, amongst the minority, more than half reported that they self-censor out of fear of reprisal from their colleagues and their employers.[23] The case is the same at the CBC, as I have already shown.

People trying to cover "the other side of the story," like veteran reporter Marianne Klowak in Winnipeg, are told there is no other side and forced out.[24] Recent reporting by journalists Dave Snow and Tara Henley has added evidence that the CBC is now what one of Henley's sources calls "an ideologically captured institution," which no longer reflects "Canadian reality."[25]

It is arguable that a doctrinaire CBC, embodying only one of the country's various political opinions, violates the very terms of the act of Parliament that gives the public broadcaster life and purpose—to "safeguard ... and strengthen" Canada's "cultural [and] social ... fabric."[26] But broadcasting acts have been flaunted ever since they were first introduced in 1932. What is more serious is the way in which the CBC has lost the country's attention. Each new era at the CBC has begun with a thrill of wild and invigorating hope. Graham Spry inaugurated the first era with a vision of radio broadcasting as "a majestic instrument of national unity," replete with "great potentialities" and "vast ... influence and significance."[27] At the beginning of the second era in the 1960s, *Seven Days*' producer Douglas Leiterman rejoiced in his "magnificent discovery" that "the power of television" could be expressed in broadcasts of such "vitality and urgency" that the whole country would feel as one.[28] Now, with the potentialities of the second era exhausted and its hope reduced to a repressive orthodoxy, a new era beckons. It will necessarily begin with a humbler and more chastened vision than the earlier eras, but it must still unfold under the sign of some galvanizing idea, like Spry's romance of communication or Leiterman's intuition of television's democratic potential. My candidate is the power of questions and their ability to open unforeseen possibilities and undisclosed futures.

So long as Canada remains divided, and without a common ground on which opposing "positions" can meet, our public and political life as a country will remain an implacable and unresolvable contest of wills—a war, like all wars, which continually feeds on itself. The law may be able to limit the potential violence of this situation, but only an authority which stands outside the terms of the conflict can restore peace. It is of this independent authority, this third term, that I have been trying to speak in this chapter. When conversation develops a spirit of its own, and the interlocutors begin to move within this spirit, such a uniting term is present. When a subject is recognized as possessing independent authority, antagonists may bow to that authority, and so to each other via their common allegiance. Tradition, as the past which is present, may reveal unperceived mutualities and correspondences. Antipathies which now seem fixed and unalterable, may show themselves as divergent branches of the same stream. The key in all cases is getting outside one's limited position and recognizing, as Gadamer says, that "the path of all knowledge runs through the question."[29]

11 | The Expropriation of Language

> Language is the house of Being. In its home man dwells. Those who think and those who create with words are the guardians of this home.
>
> —*Martin Heidegger*[1]

> Language is not just one of Man's possessions in the world; rather on it depends the fact that Man has a world at all ... Not only is the world world insofar as it comes into language, but language, too, has its real being only in the fact that the world is presented in it ... Man's being in the world is primarily linguistic.
>
> —*Hans Georg Gadamer*[2]

Many traditions agree that the origin of the world is found in language. The original peoples of Australia imagined the world as spoken into existence along the Songlines, or Dreaming Tracks, that traced the land's primeval geography. India's *Katha Upanishad* says that the seed syllable, or archetypal word, *OM* is God (*Brahman*).[3] The Hebrew scriptures portray the beginning of everything as a series of divine sayings—God said, let there be light, land, water, sky, etc. The first task of the first person God creates is naming—"The Lord God ... brought [every beast and bird] to the man to see what he would call them."[4] The Gospel of John in the Christian New Testament also begins the world with "the Word" and asserts, like the *Katha Upanishad*, that this Word is God. Creative speech, according to these traditions, is humanity's very origin, and its preeminent and defining ability. But, during the industrial age, critics, philosophers, and poets have again and again expressed the fear that language has, in Friedrich Nietzsche's words, "fallen ill" and lost its primordial power.[5] T. S. Eliot fretted that words had begun to "strain ... crack and sometimes break under [their] burden."[6] William Butler Yeats had his "mound of refuse" to lay beside Eliot's "heap of broken images."[7] Harold Innis writing a little later apprehended "a ruthless shattering of language."[8] Shakespeare and Milton did not lament in this way, nor even Wordsworth and Keats. What was new in the experience of the more modern writers was what Innis called "mechanized communication," the acceleration and multiplication of words through mass media. The words that failed Eliot had passed through too many mouths and been bent to too many other uses for him to fully trust their meaning.

Language has been instrumentalized—harnessed to so many utilitarian purposes that its creative and expressive capacity has been diminished and dimmed. Politicians have their speech-writers and public relations consultants; corporations have their carefully crafted mission statements. Each brand has its catchphrase; each song has its verbal hook. Entertainers and advertisers; announcers and administrators; journalists and educators—all jostle and compete in the economy of words, all generate their own characteristic lingos. The sheer weight of prefabricated words bears down on the space in which people might once have bent words to their own ends and, finally, saturates it. Even new countercultural idioms—at first subversive—are quickly re-domesticated, and, before long, the "underground" is playing the half-time show at the Superbowl.

This weakening of words has steadily intensified from Nietzsche's time to our own, reaching a point today where it does not seem an exaggeration to speak, as I have in my title, of an outright expropriation. A conscious curation of language must now become a primary object of the CBC. An expropriated, or coopted, language is a language in which it is impossible either to say what one means or to mean what one says. Manufactured words induce thoughts that we hardly know we're thinking. In effect, they speak for us. The instance that first made me aware of this phenomenon was the appearance around 1980 of the no longer current word, *yuppie*.

This new label was derived from the acronym for Young Urban Professional. Like many similar neologisms, it became ubiquitous and indispensable almost overnight. Though apparently nothing more than an artless bit of pop sociology, it actually helped to effect a dramatic shift in popular understanding. It did this by drawing on and transposing two earlier identifiers—hippie and yippie. The latter, no longer familiar, was drawn from the initials of the Youth International Party, a mock political party led by Abby Hoffman and Jerry Rubin. In transposing these names, the new word intimated, without having to say so, that a momentous change had taken place: The hippies and yippies, the word implicitly said, have grown up and settled back into the bourgeois groove for which they were made all along.

All they had wanted, it further intimated, was hipper forms of consumption and the chance, as the wits at the magazine *The Baffler* would later quip, to "commodify their dissent."[9]

What was at stake in the word yuppie was the meaning of the "1960s"—the scare quotes are necessary because of the avalanche of pop cultural debris under which the memory of this period has long been buried. To me, this era represented, and still represents, a turning away from thoughtless progress and toward a style of life that is fitting, proportionate, and mindful of the beauty of the many cultural styles that globalizing modernity has tended to wash away. This reversal didn't begin in 1960, or end in 1970—nor did the Sixties for that matter—but this was a time when a lot of people felt that a decisive moment had been reached. What Ivan Illich called "the possibility of making a new society, right now" was in the air.[10]

This wasn't everyone's interpretation, but the significance of this period remained, at least, an open question in the years that followed. The word yuppie decisively foreclosed the question. It pointed to the demographic explanation of the 1960s, which has since become standard. The post-war baby boom had combined with a period of unprecedented prosperity to create a period of wild social effervescence. The baby boomers had then settled down to a life of tasteful consumption, becoming Young Urban Professionals. The Sixties were revealed, in retrospect, as a consumer revolution—an aesthetic de-repression that opened new avenues of personal self-expression and new horizons of capitalist expansion.

This word, then, did a tremendous amount of ideological work, without anyone noticing, or having to notice, that this work was being done. In this way, people are induced to think thoughts that they don't know they're thinking. The word proffers itself as an explanation before the question it answers has even had a chance to form. Suddenly the world is full of yuppies—clearly, the change on which the word remarks has already taken place. This is how the magic works—I didn't initially know what the word meant, but that was clearly because I was late to the party, out of the loop, not yet hip. I have since noticed these same dynamics again and again. Another notable case was the word "app"—an abbreviation, like yuppie, and a term that also seemed to drop into everyone's vocabulary almost overnight as something already understood and taken for granted. If you had to ask—does this mean application?—you were already behind the curve—another such word. The readymade word is at hand before you even know that you are in need of it, and its sudden ubiquity makes you feel that what it describes is something unquestionable and inevitable. Of course, the world is going to be remade by apps—it's already happening.

For me, the term yuppie marked the entry into public life of demography as a pop science. Yuppies were a market segment. All the fuss about cultural revolution had been no more than a question of generational style, and of the self-assertion of a generation so big that it could push everyone else around. In saying this, I don't mean to say that population dynamics are not a primordial force, as real as fire or flood, but only to point out what was lost in this repackaging. The Sixties became the story—the story that still prevails—of the adventures of an entitled generation. The story that I favoured and felt I was living—of an apocalyptic moment urging a transformed way of life—was obscured and replaced. It would not be long before CBC executives and programmers began to talk about "target demographics." Nor would it be long before all sorts of new categories along the same lines would be invented—Generations X, then Y and Z, the millennials, and so on. A view of contemporary history as a succession of fads and generational styles, each competing with its predecessor for scarce limelight, became entrenched. So did the facile division of time into stereotypical decades that had begun with the reconstruction of the Sixties as the golden childhood of the yuppies.

The invention of the yuppie marked the coming of age of a certain style of pop sociological classification that continues to this day. It made me aware of how easily a continually restocked vocabulary of labels, epithets, and catchphrases could hide phenomena that might otherwise invite questioning. How this synthetic vocabulary comes into being is beyond my scope here, and is often obscure in any case. Ask who invented the yuppie, or the "anti-vaxxer," or who inaugurated the vast, wised-up indifference of the now ubiquitous "whatever," and you will usually find competing theories, all tenuous. What is significant, for me, is the quasi-ecology in which each finds its niche and then multiplies. In 1976, in his book *The Selfish Gene*, British biologist Richard Dawkins named the kind of tropes I'm talking about memes. Abbreviating the ancient Greek word for something imitated, *mimeme*, he claimed that units of cultural information—ideas, techniques, fashions, phrases—behave in a way analogous to his conception of "the selfish gene." They replicate, mutate, and undergo natural selection by a process "that looks like highly speeded-up genetic evolution."[11] Crucially, for Dawkins, memes direct this process, according to their "fitness," just as genes direct their own evolution and not the "lumbering robots," that is, you and I, who are their flags of convenience. I would like to acknowledge this idea but keep it at arm's length. Dawkins is clearly noticing something similar to what I've just described—ideas that propagate by a kind of contagion and possess some sort of timeliness and fittingness that makes them settle in our minds without our consciously inviting or accepting them. But to make this a natural process, occurring independently of human volition, according to the properties of the memes themselves, gives too much of the game away in advance. Terms like yuppie execute a political agenda. I do not say that this agenda is consciously planned, but I think it exists. Definite interests are reflected in the reconstruction of the 1960s as the pre-history of an emergent market segment. The word yuppie was certainly a meme in Dawkins's sense, but to make its triumph a matter of natural fitness—of ingenious adaptation to "the media ecosystem"—gives this process the force of nature. It turns cultural agencies into a meme's way of making another meme—just as Dawkins sees human beings as helpless pawns of their genes.

This is certainly the way many people do think at the moment. Talk of "the culture," "the media ecosystem," of what is "making news," etc., implies that what is talked about on the CBC, and the terms in which it's talked about, amounts to a kind of weather or wetland. Agency is erased, and responsibility with it. Subjects propose themselves. But this is exactly what I would like to contest. Unthinking use of instantaneous categories, like the yuppie, forecloses thought—it ratifies and reifies what it ought to analyze. We become "servo-mechanisms"—a favourite word of Marshall McLuhan's—of the self-replicating ideas that are at play in the media ecosystem.

The Words Do All the Work

I have claimed, by my example of the yuppie, that in adopting this usage people were subscribing to an account of the character and destination of the times through which they were living without necessarily knowing it. In the guise of an innocent description, the word did all the work for them. My example is personal, and, to that extent, arbitrary. A hundred other examples would serve as well. Speak of the Middle Ages and you immediately inscribe yourself in a historical schema, in which glorious Antiquity and its revival at the Renaissance [rebirth] are what count, with the centuries in between assigned only the status of an interruption. Speak of the Enlightenment and another whole history unfolds—one equally prejudicial to what had gone before. We cannot help inserting ourselves in a story by the terms we choose. When all agree on what the story is, we hardly need to notice our shared premises. But, when the story breaks down, and we enter an inter-regnum of competing and discordant narratives, then a peaceful *modus vivendi* can only be achieved through a conversation in which all premises are subject to scrutiny.

At the moment, we inhabit what is, in effect, a world between worlds. This can be seen as a fortunate condition—as freedom from compulsory enchantments and an opportunity for intellectual adventure. But it can only be a good fortune if it is embraced in an honest and exacting spirit. At the moment a spirit of reaction prevails on all sides. People want to go back, and this is something that those who would like to make America great *again* have in common with those who want to bring *back* truth, trusted news, and the authority of science. They want to restore and refresh modern certainties, rather than asking what lies beyond them.

In the case of language, the certainty was that our speech could be brought under control and turned into a serviceable tool. Francis Bacon, at the very beginning of what we now remember as the Scientific Revolution, portrayed the dangerous deceits of uncontrolled language as a primary obstacle to the new science he proposed. "Words beget words," Bacon said, and "pervert judgment." "Names" become "idols." They "do violence to the understanding, ... throw[ing] everything into confusion, and lead[ing] men into innumerable empty controversies and fictions."[12] When the Royal Society was formed in 1666 to pursue Bacon's vision of a new science, its members were urged by the society's charter to write with "mathematical plainness."[13] Language would be transformed into an obedient utility in the campaign "to fathom the vast depths of Nature's sea" and open its "latent schematism" to our gaze.[14] Objectivity was one of the descendants of this confident hope.

But language could not, finally, be brought under control in this way. It could not be made transparent. This is because language, in Canadian philosopher Charles Taylor's words, has a "constitutive," as well as a "designative" function.[15] In simpler words, it makes the world as well as mirrors it, creating the things we perceive as well as pointing at them. That we have a *world* at all, says Hans Georg Gadamer, in the passage I used as an epigraph to this chapter, is only because we have language in which it can show itself as a world. We are inside language, as its creatures and custodians, not outside it as its sovereign. "Mathematical plainness" cannot alter this relationship. The power of words cannot be overcome, only recognized, respected, and allowed for.

This discovery has left us in a paradoxical situation—in a world between worlds as I just said. Taylor and Gadamer, and many other philosophers of similar stature, have argued convincingly that language is the very horizon within which we live—that it is a power which creates us, even as we exercise it; an inheritance to which we must belong before it can ever belong to us. But we are also the inheritors of the Royal Society's idealistic project of a tamed and instrumentalized language. This old hope now surrounds us in hundreds of degenerated forms, which taken all together amount to a language that has been "mathematized," or alternately rendered into a code.[16]

The now ubiquitous term messaging will attest. Speech has become a kind of semaphore in which the words are flags. One sends a message and hopes that people will get it. One pushes the right buttons and hopes that the right takeaway will be conveyed. Verbal intercourse is conceived as a transaction, in which units of communication are exchanged.

These ideas have been with us since the jargon of information science began to spill over into everyday talk in the 1950s and 1960s. "Effective communication" became expert knowledge, measurable as a signal-to-noise ratio. English departments began to turn into communication departments. CBC Radio jumped on the bus in 1970, as I have related, with Information Radio. Words in this account become units of information. They lose the vibrancy that Gadamer ascribes to well-chosen words—that they "cause the whole of the language to which [they] belong to resonate."[17] Even subdivisions of language—the sentence, the paragraph, the argument—tend to disappear. Personal inflection—the unique meaning an individual may impart to a common word—is reduced. Words are used according to their value—as tokens, as signals, as stimuli—and not according to their history or their personal significance.

Plastic Words

German scholar Uwe Pörksen has spoken of words with this mathematized character as "plastic words."[18] These are words, by his definition, that have been taken from everyday speech, adopted into some science where they have been given an exact technical meaning, and then returned to common use with an added bouquet of science and expertise. Information, and its cousin, communication, are good examples. Both are old vernacular words that were drafted by science during the twentieth century and became the names of things that could be measured in bits, or bytes, or signal-to-noise ratios. This sojourn in the laboratory gave these words a new air when they were used colloquially—as for example when CBC Radio began, in 1969, to designate its new formats as Information Radio. "A failure to communicate" came to mean something more than a common-garden misunderstanding—it evoked experts with clipboards in lab coats who could advise as to just exactly how communication had failed.

All plastic words have this character—speak of *development* and throngs of economists, like angels with trumpets, attend the word; *sexuality* evokes supervising sexologists; *role* is redolent of sociological role theory, and so on. Pörksen listed some fifty words of this character—*energy* and *strategy*, *resource* and *system* are other examples—and a few new ones have expanded this roster since he wrote. What unites these words is their connotation of authoritative expertise and universal application. But these words are in fact imposters—all hat and no cattle, as the saying goes. They once denoted something precise—within the sciences that outfitted them—but in ordinary speech, they denote nothing at all. All that remains is connotation—an endlessly swelling feeling of significance. "A plastic word is like a stone thrown into a conversation," says Ivan Illich, who developed the idea with Uwe Pörksen. "It makes waves but it doesn't hit anything."[19]

In ordinary speech, words become precise by the way in which they're used. The many meanings a word might have are refined and sharpened by conversation and context. The plastic words don't work in this way. Being all aura, or significance without reference, they cannot be adapted for ordinary use. Rather, they recruit us and are, in this sense, a species of anti-word. They point into an institutional abyss where they can be "strategically" deployed according to that institution's purposes. Not only have they become useless as vernacular words, they have become actively harmful by the way in which they displace other more fitting words.

Plastic words have an imperative rather than a descriptive character. Each word is a mandate, an instruction, or a command. You ought to interrogate your sexuality, be cognizant of your role, and be attentive to whether you're communicating effectively. Like a light that "shines everywhere," Pörksen says, such words lack any shadow, nuance, or possibility of personal inflection.[20] They are modular, like plastic Lego blocks, made for combination. They are inert—like wads of cotton batting. And because they claim universal validity—a process is a process, a system a system, the same everywhere forever—they disregard history and culture.

Plastic words reorganize experience. Consider one that Pörksen didn't mention: stress. Before Hungarian-Canadian endocrinologist Hans Selye got hold of this word, in the 1930s and 1940s, it had quite a restricted reference—airplane wings experienced stress; people did not. Selye took the word into his lab, and it returned, by and by, into everyday talk as a term of such astonishing generality that it no longer pointed at anything in particular.[21] Anger or grief, haste or worry, too much work or too little—it's all stress. Scores of apt and colourful words have been eclipsed by this quasi-scientific pretender. But stress has the power. Its scientific aura creates respect and commands resources—you can get a "stress leave"—and so people shape and reshape their experience in its image. Reality forms in the field of the model.

Some friends of mine in Germany once did a study of this phenomenon of scientific words in everyday talk, focusing on what they called "the pop gene," or the gene as it turns up in the talk of butchers, bakers, and candlestick makers in small-town Germany.[22] Within the science of genetics, the term *gene* is a convenience—a shorthand way of conveying an extraordinarily complex and not always clearly understood terrain. The scientists who use the word know what they are saying and what they are not saying. They know particularly the crucial difference between cause and correlation—the difference between implicating a certain sequence of nucleotides in some subsequent condition and demonstrating that the one, in some simple way, caused the other. On the street, this distinction is lost, as my friends discovered. The gene ceases to be a restricted and qualified concept and becomes a thing with quasi-magical properties. It acquires what Alfred North Whitehead calls "misplaced concreteness"—his name for the fallacy of taking abstract ideas for perceptible realities.[23] One can think by analogy of the way in which COVID-19 became a thing through the work of cartoonists, illustrators, and animators—an invisible, infinitesimally tiny object, whose existence is inseparable from the theories and the visualization techniques that make it manifest, is transformed into the malicious spikey imp that everyone can now picture. No wonder people thought that masks would easily repel it. The point is that when technical and scientific terms enter ordinary speech, two spheres are mingled. This intercourse is not harmful to science, which emerges from the encounter unscathed and still knowing what a gene is and what it isn't. But the scientific word, let loose in the vernacular as if it were an everyday word, becomes pure myth and a potent source of confusion.

Words of this kind are deprived of the quality that Gadamer calls resonance—the ability to induce sympathetic vibrations in other words and even in the language as a whole. This effect can be seen in music, where the playing of a note causes the entire "overtone series" of that note to sound. These overtones are heard as part of the "fundamental" pitch. A single tone turns out, on analysis, to be a whole cluster of tones. Similarly, there are stringed instruments from around the world that have what are called sympathetic strings—strings whose primary purpose is to resonate with the main strings. But, in language, such resonance is damped when words are institutionalized. Their communion with other words is interrupted. A plastic word, like a baleful god, can only be obeyed.

Uwe Pörksen, writing in the 1980s, already feared that language was becoming "mathematized," that it was becoming a kind of source code from which social instructions could be "executed."[24] Subsequent developments have only strengthened his claim that language is being simplified and words turned from living things into inanimate diagrams. One such development is datafication, the rendering of the world into ever smaller and ever more precise bits of information. In sport, no gesture can now escape measurement—the velocity of the pitch, the launch angle of the home run, etc. In medicine, no condition can go undiagnosed. In public discussion, no predicament can avoid classification and assignment to a place in the litany of crisis. And, if it hasn't been datafied yet, it soon will be. Hardly a day passes without complaints about the insufficiency of our data banks. "In the dark," reads a typical headline in *The Globe and Mail*, "the cost of Canada's data deficit." The story goes on to warn of all the things we've neglected to count—"from divorce rates to driving patterns to labour market trends"—and warns of the trouble we'll be in if this carelessness continues.[25] This mania inevitably casts its shadow on language. If all is data, then language too must be data.

A second development is the gradual erasure of language's defining boundaries. As early as 1943, in a lecture series at Trinity College, Dublin, entitled "What Is Life?," physicist Erwin Schrödinger put forward the suggestion that human heredity—the genetic code, we now say—could be understood as a form of text.[26] Ten years later, James Watson and Francis Crick revealed the letters in which this text is written with their model of the DNA molecule—"the language of life" was revealed. At the same time, the infant science of cybernetics was exploring the idea that the world is best understood as made up of informational patterns rather than embodied forms. "We are but whirlpools in a stream of ever flowing water," wrote mathematician Norbert Wiener, the man who named the new science, "not stuff that abides, but patterns that perpetuate themselves."[27] All across the sciences, life and language were dissolving as stable, integral, privileged entities. Language lost its boundary—it became the waggle dance of the bee and the song of the white-throated sparrow, the nucleotide letters in which the genome is written and the speech of plants conversing along their mycorrhizal networks.[28] Everything became language, and language became code. René Descartes's old dream of a "universal mathematics"—*mathesis universalis*—a mathematics able to "contain everything" had been realized.[29] As a mere instance of a universal, biological coding, language lost its specificity, its link to particular peoples, places, and times.

* * *

At the head of this chapter, I placed mottoes from Martin Heidegger and his student Hans Georg Gadamer: Heidegger speaks of language as humanity's home and dwelling place; Gadamer makes it the condition of our having a world at all. Both suppose that language is integral—that it has properties as a whole that do not belong to any of its components—and that it is historical—understandable only with reference to all that has already been in language.

In the preceding pages, I have laid out some of the challenges I see to this view of language as our defining possession and chief treasure. These include words twisted and spun to implicit and unmarked ideological purposes, as it is with pop sociological terms like *yuppie*; language infiltrated by glamorous migrants from technoscience that become empty husks on the common tongue; and, finally, language "mathematized" and reduced to a communication code by which information is exchanged. Taken together these conditions amount to an expropriation of language—the loss of what is proper to people as members of the species that Charles Taylor calls "the language animal."[30] I do not mean, of course, that there is no resistance to this deprivation, no counter-current to this tendency. I want only to indicate what might be called the default mode of contemporary discourse—the language at hand for automatic and unreflecting use. In this language, people can no longer mean what they say or say what they mean because the available words and phrases execute agendas to which they hardly know they are consenting.

Two things need to be restored. The first is a certain balance or proportion between what people can say for themselves and what arrives ready-to-wear from Big Language. Some part of language has always previously been homemade, particular to a family, a place, or a region—its words passed on to us by people we have known from traditions to which we have belonged. Now manufactured language threatens to inundate this vernacular remnant. This is the first issue I see for public broadcasting: Is there some combination of criticism and example by which it can at least restrain the flood of over-programmed words and allow its listeners a space in which to shape words for their own purposes?

The second thing that needs restoration is the common tongue, the public language with which people must try to understand one another and get a grip on the questions that are facing them. At present, this task is not even recognized as a task, let alone an obligation. Language is the first and most fundamental of commons—as vitally necessary to our subsistence as air, soil, or water—but its care and conservation appear nowhere among the CBC's stated objectives. Such care will be vital if the CBC is ever to create the new public forum that Canada now needs. Only a language that is recognized as a common care and a common possession will have the power to align our warring tribes within recognized courtesies, histories, and fields of meaning. Without such a language, people will continue to speak past one another. Whether we're talking about the scholar at home in the rare air of high post-modernism or the inarticulate youth who occupies the verbal space that American writer Clark Whelton calls "vagueness"—who's like, you know, whatever, and stuff—the problem is the same: Neither can be understood by anyone outside their particular milieu.[31] Here I think the CBC needs to provide an example—by consciously speaking what might, at least, *become* a common language. The potential of language—to mean what the participants in a conversation hope to say—must be recovered by holding at bay the overdetermined and ideologically saturated words that I have described.

In proposing a more conscious curation of language, I'm not suggesting that the CBC should begin to put on airs or adopt more high-flown ways of speaking. Nor do I imagine the creation of some sort of *Académie Canadienne* to police the boundaries of acceptable discourse and keep pop sociology and plastic words off the public airwaves. If everyone is suddenly talking about yuppies, or millennials, or whatever the category *du jour* may be, the CBC can hardly hold its nose and refuse to join the conversation. Rather, I'm recommending two things, which I think are perfectly achievable. The first is critical awareness—the recognition that discourse is neither innocent nor neutral. To speak of yuppies, as if they were a natural kind like trees or flowers, is to ratify the version of history that is encoded in this invented name. CBC broadcasters may not by themselves be able to prevent the spread of the ideas insinuated by such words, but they can at least become conscious of the *un*conscious ideological dynamics in which they are implicated.

The second achievable end is to identify the margin of freedom that the CBC still retains—even in the midst of constraints of every kind. This margin may be small, but I do not believe it is non-existent. If only one step is available, take that step. One step and the world changes—perhaps in the new configuration a second step will be possible. Change needn't occur through master plans, vision statements, and the whole apparatus of outsized solutions to outsized problems. It can occur by changing something within reach: In this case, by practising critical self-awareness and discerning attention to the available margin of freedom.

This is a question of vocation more than of regulation. Language requires conservation, even if it is not normally included among the things that environmentalists hope to save. A ruined language belongs to the "environmental crisis" just as surely as a warming climate, or the plight of the rusty-patched bumble bee. We will not heal a world we have lost the ability to name. The same is true in the political arena. We will not restore amity between enemies unless we provide them with a common language and then hold them to its affordances. Both assignments ought to become part of the new vocation that the CBC now plainly needs.

12 | Thinking Must Always Begin Afresh

> To think and be fully alive are the same, and this implies that thinking must always begin afresh; it is an activity that accompanies living.
>
> —*Hannah Arendt*[1]

In 1967 and 1968, when he was near the end of his life, German philosopher Theodor Adorno found himself beset by the student movement that was then bent on revolutionizing German universities. Adorno had been a man of the political left throughout his life, and he shared many of the student movement's aims, including university reform, but he became a target of student protest nonetheless. In classroom discussion of his book *Negative Dialectics*, students, including visiting American Angela Davis, objected to the work on the grounds that he had failed the first test of "all critical thought"—that it "must adopt the standpoint of the socially oppressed." Adorno replied that his book was concerned rather with "the dissolution of standpoint thinking itself"—a position that his students felt was intolerably academic and insufficiently *engagé*.[2] The trouble escalated in 1968 when the Institute for Social Research, of which Adorno had been one of the founders, was "occupied" by student protestors. Adorno along with the other directors summoned the Frankfurt police, and one of his graduate students was charged. The following year, when Adorno began his course, "Introduction to Dialectical Thinking," he was immediately approached at the podium by two tall men from the direct-action wing—the so-called leather jacket faction—of the *Sozialistischer Deutscher Studentenbund*. They demanded that he engage in self-criticism for his role in ending the occupation of the institute. Then, according to Adorno's biographer, "he was surrounded on the platform by three women students who scattered tulip and rose petals over him. They then bared their breasts and tried to approach him while performing an erotic pantomime."[3] Adorno, embarrassed, walked out of the class.

Adorno objected to the New Left's "propaganda of deeds"—its taste for self-righteous self-dramatization—its "ridiculous" theatre of "building barricades … against those who administer the bomb."[4] He never renounced his support for most of the movement's aims, but he would not submit to the tactics of these "brownshirts in jeans," as he once called them.[5] The students for their part mocked his supposed scholasticism, unfurling a banner when he lectured on Goethe's *Iphigenia at Tauris* at the Free University in Berlin that read "Berlin's left-wing fascists greet Teddy the classicist."[6] As far as they were concerned, he was fiddling while Rome burned, and they accused him of fatalism and "resignation."

Adorno was disconcerted by these attacks. "Valid student claims and dubious actions are all so mixed up together," he wrote to his friend and colleague Herbert Marcuse, "that sensible thought [is] scarcely possible anymore."[7] He responded to his critics, nevertheless, in a radio talk that he titled "Resignation."[8] In his talk, Adorno reversed the charges against him. It was his student critics, he said, who had resigned. They had resigned from the task of "open thinking" by engaging in what he called "pseudo activity" or "actionism." They were trying to "rescue enclaves of immediacy in the midst of a thoroughly mediated and rigidified society" because they were unable to accept the impossibility of remedying this situation through direct action. "One clings to action because of the impossibility of action," he wrote, and because "thought … not immediately accompanied by instructions for action" provokes anxiety. How much easier for the student activists to rage against people like him who try to expose "the pseudo-reality within which actionism moves" than to face up to their situation.[9] But this "prohibition of thinking" is fatal, he argued, because there are situations in which existence is so completely "mediated" and "rigidified" that "only thinking [can] find an exit."[10] And this thinking must be "a thinking whose results are not stipulated, as is so often the case in discussions in which it is already settled who should be right." Thinking, he concluded, always "points beyond itself." Such thinking is "happiness even when it defines unhappiness" because it "reaches into the universal unhappiness" and "enunciates" it. Thought is a power that is, by its nature, free and "whoever does not let it atrophy has not resigned."

Reading Adorno's words recently, and imagining the old man's distress in the face of his confident tormentors, I thought of economist and communications scholar Harold Innis, who also mounted a defence of thinking against "actionism" in the Canada of the 1930s. In the first years of the Depression, a number of Innis's colleagues at the University of Toronto expressed their support for socialism, and for the Cooperative Commonwealth Federation, the ancestor of today's New Democratic Party. Innis demurred and questioned whether his politically engaged colleagues were really wise enough to design and install "the new social order" called for in the Regina Manifesto, the CCF's founding document. "We know very little about the solution to so-called economic problems," he wrote to his historian friend Arthur Lower, "and there is little point in concealing our ignorance by loud talk."[11] "I am sufficiently humble in the face of the extreme complexity of my subject," he said on another occasion, "to know that I am not competent to understand the problem, much less the solution." Innis's reticence was taken by his more activist co-workers as complicity in the status quo. The best known of them, historian Frank Underhill, said scornfully that economists like Innis were no better than "the garage mechanics of capitalism."[12] Even Innis's generally admiring successors have tended to be embarrassed by his stance.[13] Today, to me at least, Innis's position appears in a different light.

Innis took seriously the great difficulty presented by thinking—that it must somehow find a place to stand outside the conditions that have formed it in the first place. In his communication studies, as we have seen, this view led him to the idea of counter-vailing or counter-posed media. The only hope of withstanding the spatializing effects of contemporary media, he argued, lies in our retaining or redeploying media that amplify the opposing dimensions of time, tradition, and orality. In the face of the assured activism of his fellow scholars, he insisted that the university could only fulfill its vocation as a space for thought by standing back from political urgencies and carefully avoiding premature commitments and *a priori* judgments. On occasion, he even dared to give the usually derogatory image of the Ivory Tower a positive valence. "The university is essentially an ivory tower," he wrote, "in which courage can be mustered to attack any concept which threatens to be a monopoly."[14] For Innis, a monopoly, or as he usually said, a monopoly of knowledge, was a thought-style so dominant that it entrained all minds and set them running in predetermined "grooves."[15] Adorno meant the same thing, I think, in speaking of a society so "thoroughly mediated and rigidified" that the only hope of escaping its gravity lay in the power of independent thought. Canada in the 1930s was a very different place than Germany in the 1960s, and Innis's sense of the limits of his concepts rested on different grounds than Adorno's efforts to open new paths for philosophy, but the two men still had a crucial idea in common: that thought exceeds all predeterminations and stipulations as to "who should be right" and must be exercised in a space free from the exigencies of immediate action.

The predicaments faced in their time by Adorno and Innis also make me think of our situation today. When a CBC News Network executive banned Ontario's former chief medical officer of Health, Dr. Richard Schabas, from the CBC in March 2020, on the grounds that his criticism of pandemic control measures was tantamount to "climate change denial," he acted for a corporation that now countenances only correct opinions.[16] In this sense what Adorno called a "prohibition of thinking" is currently in effect at the CBC as well. The stakes seem too high, the need for action too urgent, the correct view of the situation too obvious, for anyone to waste time indulging in what Adorno calls "open thinking."

A Place to Think

I believe, on the contrary, that in our present circumstances, open thinking constitutes the highest and best use of the CBC's public monies and the most cogent contemporary version of its legislative mandate. If the Conservative win the election that is in progress as this book goes to press, the CBC will face a government that wants the CBC to become "a true public service broadcaster" whose offerings do not "overlap ... or compete with private sector equivalents."[17] This is a new version of a venerable complaint against the CBC—that it unfairly competes with private broadcasters, stealing scarce advertising dollars and using its public subsidy to produce programs that the privates could supply just as well.[18] Since 1958, when regulatory authority was removed from the CBC, Conservative governments have tended, once elected, to back away from the anticipated trouble of actually acting on this complaint. But, this time around, there are cogent reasons for supposing that a Conservative government might be prepared to make the attempt to restrict the CBC to "public service."[19] If it does, a question is going to arise about what exactly constitutes public service under current conditions. So far, the Conservative Party has not even posed this question. Slogans, like "defund the CBC" bring wild applause at rallies, but no positive program for public broadcasting is put forward.

A place to think is an answer to this question. With its universities transformed into "political monoliths," and its politics tied up in sterile hostilities, Canada plainly needs such a place.[20] To provide it would, therefore, be an undeniable public service, and one that private, for-profit media are quite unlikely to perform. But what, then, would be the character of such a place? I intend the word thinking in the sense in which Adorno and Innis used it—to name a faculty able to resist the pressure of what Adorno called "actionism" and patiently search for paths where none presently appear. I have argued in these pages that the modern era, as well as the CBC's populist adaptation to it, is now at an end—its former certainties stymied and contested. This is a situation that manifestly calls for thinking, and for thinking, moreover, with the quality that Innis unaffectedly called "humility"—the humility of recognizing, as he says, that we are scarcely "competent to understand the problem, let alone the solution." The CBC was born on the cusp of a great era of progress—an era in which the language of "destiny" and "national ideals" came readily to hand for Graham Spry and R. B. Bennett alike. This era continued, in a new key, in the populist visions of Patrick Watson, Douglas Leiterman, and the heroes of the radio revolution. Today, the very meaning of progress is obscure—the way forward blocked by an accumulating jumble of competing definitions and contending thought styles—the very idea of solutions mocked by the amorphous and illimitable complexity of the problems—no one quite sure whether heaven or hell waits round the corner. Even more than when Adorno spoke the words in 1968, we are now in a situation in which "only thinking can find an exit."

Such thinking, in my view, would have two prime characteristics. The first is that it would be unfinished and unfinishable—a conversation with more always to come, rather than a quest for some decisive last word. *Inconclusive* was the word Innis used in his praise of the "oral tradition."[21] Philosopher Hannah Arendt (1906–1975) said that "thinking must always begin afresh." She argued that once philosophy had given up its claim to know exactly and authoritatively how things stand—something she believed had happened in her time—thought could no longer aspire to some definitive conclusion, or final resting place. It must be, she wrote, like the loom of Penelope, in the ancient Greek epic *The Odyssey*, on which what is woven by day is unwoven again at night.[22] This, for Arendt, was not a figure of futility or defeat, but rather a bracing and enlivening image of the not-yet and the still-to-come that thinking endlessly pursues. "To think and be fully alive are the same," she wrote, because one must endlessly resist the ways in which the already thought can paralyze the need for further thinking.[23]

The second characteristic that I want to emphasize is thinking's endless difference with itself. I recognize that thinking has a participatory and intuitive mode, as well as a more detached and analytic manner, and I don't mean to prefer one over the other, but what I want to point up here is thinking's need to stand back and slacken the power of identification—the making of identities which pre-empt further questions because they have come to rest, complete and self-contained, in themselves. "If thinking is to be true … today," Adorno wrote, "… it must … be thinking *against itself*."[24] Such thinking must accept its inevitable plurality, rather than trying to reimpose the unity it imagines it has lost.[25] Arendt expressed a similar idea. "The essence of thought," she wrote, is to be "two-in-one," and therefore always in conversation with itself.[26] This doubleness, or plurality, is particularly crucial in the contemporary situation of failed certainties, warring ideologies, and crossed purposes. Thinking's task, under these circumstances, is to instill questionability, to let light and air into closed rooms, and to break apart confident and mobilized identities. And, for this, thought's restless difference with itself is indispensable.

However, if thinking is to become one of the CBC's new vocations, four formidable obstacles will first have to be faced and removed. Each, as I will show, inhibits thought. The first is sentimentality. The second is what literary critic Peter Brooks calls "the narrative takeover of reality," or the reduction of all modes of existence to story.[27] The third is the pursuit and enforcement of identity as the primary business of political and social life. And the fourth, and last, is the generation of an atmosphere of perpetual emergency, an atmosphere so intense that thought comes to seem an unaffordable luxury. I will take each in turn.

Sentimentality, the dictionary will tell you, denotes an excessive, or perhaps an unsuitable expression of feeling. Czech-French novelist Milan Kundera helpfully substitutes the German word "kitsch" and defines it as "the second tear." "Kitsch," he writes, "causes two tears to flow in quick succession. The first tear says: How nice to see children running on the grass! The second tear says: How nice to be moved, together with all mankind, by children running on the grass!"[28] Sentimentality pretties things up, allowing people to believe that things are better than they actually are, and that we—the righteous, caring people—are better than we are. The past is met with endless apology—we would not have done such things. "Safe spaces" are created to avoid offence, triggering, or retraumatization. Mediocre achievements are overpraised—the incredible new album, the awesome new song. The execution of ordinary, well-compensated duties evokes excessive gratitude—the newsreader thanked so much for reciting his five minutes of script, the neon heart in the window for our front-line workers. Ambitions become dreams; celebrities become icons; motives lose their shadow. Grubby necessity is scrubbed out of the picture and replaced by heroic and disinterested care. This is now so pervasive, on the CBC as elsewhere, that it's barely noticeable. But what has happened is that a fatal falsification has taken place. Performed feeling begins to overshadow real feeling; unwarranted superlatives begin to deprive real excellence of its proper measure; emotion swamps judgment. Perspective and proportion are lost. All this is fatal to thinking because thinking must be able to bracket easy and instantaneous emotions, it must be able to withdraw and wait rather than enact an expected response, and it must be able to confront a reality that has not been ideologically embellished (at least as far as that's possible). Where sentimentality prevails, thinking will appear as a downer, a kind of willful alienation from the all-important task of sorting out how one feels.

A second impediment to thought is the widespread idea that all mental activity can and should be reduced to a story. I discussed this issue earlier in relation to Thomas King's assertion, in his 2003 Massey lectures, that "The truth about stories is that *that's all that we are*" [my italics].[29] I argued that stories are, in fact, only half of what we are, the other half being our ability to go against the grain of the narrative by thinking about, and against stories. Here I'd like to take an example of the power of stories that was something of a revelation to me. It comes from McGill anthropologist Ronald Niezen's book, *Truth and Indignation*, an ethnographic study of Canada's Truth and Reconciliation Commission (TRC) on residential schools. This book is a careful investigation of the tension between "historical truth" and the universal desire for a consistent and compelling narrative. Niezen writes as someone fully and sympathetically aware of the many harms that were done in residential schools, but also as someone who couldn't help noticing the ways in which the Commission on Residential Schools tended to produce a stereotypical narrative to which everyone contributing was encouraged to conform.

This was accomplished in several ways. First, there was the implicit command to believe the victim—already well-established by the time the TRC began its hearings in 2007. "What witnesses convey[ed] in their narratives," Niezen writes was "commonly understood to be unadulterated, veridical reports of lived experience, not instrumentally limited reports that [were] subject to the subjectivity and omissions of memory."[30] One overlooked the fact, in other words, that the witnesses' testimonies were "rhetorically organized." Second, the testimony of the priests and religious who worked in the schools was sharply limited. The churches were expected to listen not speak, and to this end "church listening areas" were provided at the hearings where "survivors could sit down individually with a church representative of their choice and hear a private personal apology."[31] Finally, there was the performative quality of the TRC hearings. The Commission, Niezen says, was "seeking historical truth" in conditions in which the testimony it heard had "the heightened quality of public performance … in which approval is garnered for emotional poignancy, moral clarity, and narrative continuity"—conditions, Niezin adds, which are "inimical to 'historical truth.'"

This brings me to the incident I want to relate, which, according to Niezen, was the only occasion on which he saw "a church representative disrupt the narrative of contrition." I will quote the story in full:

> When Brother [Tom] Cavanaugh, representing the Oblate order, made a presentation to the Commission at an event called "Expressions of Reconciliation," he came, or so it later seemed, with the goal of setting the record straight. Before an audience of some 800 people, consisting mostly of survivors and their family and friends, in a large meeting room of the lavishly appointed Fairmount Empress Hotel in Victoria, British Columbia, he expressed views that differed sharply from the usual paradigm of church contrition. Brother Cavanaugh spoke of his experience (which he introduced as "my truth regarding my experience") as a young man, 21 years old, appointed as a supervisor of the senior boys in Grades 5 to 8 in the Christie residential school at Kakawis on Meares Island (off the west coast of Vancouver Island), which he remembered as a place of love and dedication to learning, with the Sisters of the Immaculate Heart bearing a great responsibility for the education and welfare of the students, together with five Oblate Fathers and Brothers as well as eight First Nations support staff. When he said in passing that the 120 children in the school were "sent by their parents," there was a murmur from the audience. He went on, "Was it a perfect situation? No," and followed up with a statement that seemed to set things off: "There didn't seem to be any other viable alternative in providing a good education for so many children who lived in relatively small isolated communities."
>
> At this point there was a shout from the audience: "Truth! Tell the truth! You're not telling the truth!" From somewhere else in the room there was a loud wail, followed by agonized weeping.
>
> Cavanaugh persisted through the interruption: "the Native staff who were related to a number of the children, along with the other staff, I felt, provided a good education, as well as excellent care and guidance."
>
> "Tell the truth! Shame on you! We never sent our children to residential school!"
>
> "Parents were encouraged to visit the school and rooms were available, if they wished to stay overnight to be with their children."
>
> "Tell the truth!"[32]

This story made quite an impression on me—not because I think Brother Cavanaugh's witness has any more claim to being "veridical" than competing testimonies, but because of the immediate accusation that, in deviating from the established template, he had obviously failed to "tell the truth." Among the disagreeing details are the lodgings for visiting parents, the active role of native staff at the school, the agreement of parents in sending their children to the school, and the "love and dedication" of the school's faculty. These details, accordingly, were not treated as anomalies worthy of consideration, or further investigation—they were treated as falsehoods, things that, by definition, could not be true.

The reception Brother Cavanaugh received is similar to that accorded to Senator Lynn Beyak who spoke about residential schools in 2017, saying that even though she did not want to "minimize the sufferings of the victims," she did want to say that there were also some "kindly and well-intentioned people" involved in residential schooling."[33] This touched off a furor, and her statement was said by one member of Parliament to be tantamount to an endorsement of the Holocaust. Beyak was twice suspended from the Senate and made to undergo "anti-racism training." In early 2021, facing final expulsion from the Senate, she retired.

The question that I think is raised by both cases is not whether Beyak or Cavanaugh are right, but whether one is allowed to hold an opinion that varies from the established "truth" to which Brother Cavanaugh's audience tried to hold him. This is neither a new nor a trivial problem. Thomas Aquinas (1225–1274), generally thought to be a good and peaceable man by those who knew him, held that heresy should be punishable by death because of the harm done to the community of the faithful by those who "corrupt the faith" by "devising false or new opinions."[34] The relevant section of his *Summa Theologica* could easily be used to indict Brother Cavanaugh today. An image of residential schooling as an entirely vicious and ill-intentioned genocide has become an enforceable dogma in Canadian public life, as ex-Senator Beyak's fate shows. The end this story serves is arguably just as admirable as the end which Aquinas sought to preserve at all costs—the unity of the church. Reconciliation, healing, and decolonization are all seen to depend on the careful maintenance of this narrative, and so it may not be touched or questioned in any way. The story has become compulsory, and one cannot interfere with it without being thought hard-hearted, morally obtuse, and, like Lynn Beyak, racist.

Such compulsory stories are incompatible with thinking. Thinking, to deserve that name, must be free and uninhibited by any demand that it produce constructive or consoling narratives. It cannot be made to bow to "settled science" or official history, nor allow its boundaries to be controlled by the denialism and disinformation police. Edifying stories that have become mandatory now surround us on all sides. "The science" one was expected to "follow" during the recent pandemic was not the disinterested product of free inquiry—it was a militant, mobilized, and interested opinion prepared to "take down" whoever disagreed with it.[35] The "truth" one is required to believe about residential schools is not the result of open-ended historical investigation but of a process of narrative coalescence whose ends are political and therapeutic. "Which side are you on?" now seems to be a more important question than, what is the case? We are back to Adorno's "propaganda of deeds." Thinking, which is always rethinking, hasn't a chance.

Identity

The third enemy of open and unforced thinking that I want to consider is identity politics. Let me again begin with a telling incident. Back in 2017, Canadian writer Hal Niedzviecki offered his opinion in the Writers' Union of Canada journal that "anyone, anywhere, should be encouraged to imagine other peoples, other cultures, other identities" and backed up his advice with the jocular suggestion that there should be a prize rather than a penalty for cultural appropriation.[36] The sky immediately fell on him. He was forced off the magazine's editorial board and accused of "kick[ing] indigenous writers when they're already down" by claiming to be able to "tell their stories better than they can."[37] Niedzviecki had shown no inclination to tell Indigenous stories, let alone presumed that he could "tell them better." He seemed only to be defending and celebrating the freedom of the imagination, but he had evidently crossed a boundary without showing his papers. The message was plain: The only people with the right to speak about a given identity are those who belong to and own that identity. Their reality, their truth, and their experience are what they say they are, and nothing else.

Identity politics, of which Niedzviecki was a victim, currently has the force of an edict, able to draw boundaries, distribute privileges, and determine not just who will speak but what they will be allowed to say. Assertions of identity create an atmosphere of intimidation in which most will be silenced, and those who dare to defy censure will often muster the courage to do so by becoming belligerent themselves. Conversation will be blocked in either case. Identity is an essence—something imperative, irreducible, and unconditioned. It ends all inquiry and answers all questions. Possessing it, one needs only to defend its prerogatives and extend its scope. Identity, as its adjectival form "identical" makes clearer, refers to something that is the same as itself—something finished and fully known, without contradiction, remainder, or more to come. This makes it more or less the exact opposite of the self-estranged, "two in one," in endless conversation with itself, by which Arendt defines thinking.

Here a strange irony presents itself—the apostles of identity often declare themselves, at the same time, as the enemies of essentialism, creating an apparent contradiction. On the one hand, anyone who is not white can be said to have been "racialized"—a term which seems to say that race is no more than a construct and corresponds with no essential quality. On the other, identity is treated as an inexhaustibly productive essence, something so *proper* to its bearer that it must not be *appropriated* or even imagined by outsiders, like Hal Niedzviecki. The contradiction makes the discourse of which it's a part invulnerable—its anti-essentialist, except when it isn't. All robust ideologies have this flexible, shape-shifting character, but it's a tyrannical power nonetheless. What ought to be questioned, and can only yield a way forward if it is questioned, is instead set aside and made the exclusive property of whichever identity is implicated. Thinking is impossible where certain groups and individuals are empowered to authoritatively set and police the terms of public discussion.

The last of my four hindrances to thinking is the widespread feeling that we live in a crisis, or emergency, so profound that there is no time to think. It used to be the case that the word crisis referred to the moment at which a sickness turned decisively toward death or recovery. It described a turning point at which the illness was bound to resolve one way or the other. A crisis, by definition, couldn't last. Current usage is the exact opposite. Crisis is now so normal that one can hardly think of an institution or condition to which the term is not regularly applied. Resolution is no longer expected, and perhaps not even desired, since the care, feeding, and management of crisis is, by now, the very raison d'être, and livelihood, of so many organizations. This state of normalized emergency generates, along with anxiety and foreboding, a feeling that time is unavailable, that it's always already too late. It becomes safer and more prudent to live in the future rather than in the present. Speculative models govern social decisions—as in the case of the recent pandemic, when policy was set on the basis of what turned out to be wildly incorrect projections.[38] People regularly act on the basis of their risk rather than their actual experience. Inhospitable and threatening futures crowd into the present and preoccupy it. Time to think becomes irresponsible delay.

The difficulty here is that we are in a situation in which, as Adorno said, "only thinking can find an exit." This is true regardless of how acute our encircling crises may be. Harold Innis's colleagues were convinced in the 1930s that there was no time for the thoughtful and sequestered inquiry to which Innis was committed. Adorno's student antagonists thought the same in 1968. But does it make sense to act when you don't know what you're doing? Isn't this just the absurdity of "build[ing] barricades against … the bomb" that Adorno deplored? He found the "actionism" of his critics to be nothing more than a theatre of reassurance intended to allay anxiety and to postpone the one thing needful, which was patient thinking. Innis called his politically involved fellow scholars "hot gospellers of truth" and "travelling salesmen peddling nostrums."[39] I think the case is the same today. In a situation of confusion and worsening conflict, I see no alternative to a reconsideration of the terms that divide and confuse us.

Thinking, even in impossible circumstances, asserts its freedom. When all manner of *crises* are brought before us—and brought before us in terms so fraught that they inhibit understanding of how they came to be crises in the first place—only thinking can untangle this self-tightening knot. But thinking requires disengagement, a capacity to sustain self-division, and what Innis called an Ivory Tower—a space "in which courage can be mustered to attack any concept that threatens to become a monopoly." Emergency, with its call, at all costs, to do something, is its adversary. It follows that the CBC will only be able to become the place where Canada thinks when it begins to question the reflexive rhetoric of crisis.

New Habits of Thought

I have identified four stumbling blocks in the way of thinking: a panicked atmosphere, closed identities, the tyranny of narrative, and epidemic sentimentality. Where these powers predominate, thinking is passive and runs only in approved channels. Deviance can be actively dangerous to livelihood and reputation. Settled science, authorized history, and zealous identity and misinformation policing create rhetorical minefields. And yet, as I've tried to show, what our current situation most urgently demands is active and open thinking. How then is the CBC to address this urgency?

I should say first that I don't imagine the corporation suddenly transformed into a full-time academic seminar or debating society. I am talking rather about new habits of thought—habits which might begin in anyone and be expressed in the production of any type of program. Loosen, if only a little, what William Blake called our "mind-forged manacles."[40] Open the "prisons … we choose to live inside," as Doris Lessing titled her 1985 Massey Lectures. Begin to think differently.

The CBC currently conceives itself in positive terms—as a place of uplift, belonging, and correct theory. Advocacy and allyship, supportiveness and sensitivity are emphasized. The solidarity of broadcaster and listener is continually reinforced by the use of personal pronouns and definite articles—*your* weekend, *our* society, *the* science, and so on. *Our* take on *the* world is not put into question. What I am proposing might be described as the development of a negative space—though perhaps hypothetical, potential, or even playful might give my idea a more agreeable air. What I mean, in any case, is a space that is not pre-occupied, pre-defined, or pre-judged. Its negativity would lie in its suspending or holding at bay the immense positive pressure of the taken-for-granted. Poet John Keats called this suspension "negative capability"—the capacity, he said, to sustain "uncertainties, Mysteries, [and] doubts," without any premature rush to judgment.[41]

"Method of investigation," the philosopher Simone Weil once jotted in her notebook, "as soon as we have thought something, try to see in what way the contrary is true."[42] Another philosopher, Alfred North Whitehead, late in his life, told an interviewer that, "There are no whole truths." "All truths are half-truths," he continued. "It is trying to treat them as whole truths that plays the devil."[43] Hannah Arendt's insight that what is "trans-acted" in thought is "my difference from myself" also chimes with Weil's "method."[44] All three say that our thinking is inevitably incomplete—that it must continually "begin afresh" and continuously reform its own tendency to rigidify and become self-enclosed. It must reach for, and wait for, the sense of direction that sometimes takes hold of those who trust themselves to the mutuality of open-ended conversation. It cannot be for or against what it has yet to understand.

Such a stance—or non-stance, as Adorno's student critics claimed—is difficult. In "thinking against itself" it cuts the ground from under its own feet. It cannot be satisfied with solacing, ready-to-hand pieties and platitudes. But it also offers a profound sense of encouragement, inspiration, and intellectual adventure. Exhausted words can be refreshed, worn-out certainties opened to new interpretation, and an open future discovered in an unlocked past. An age has died—the modern age according to those who speak of post-modernity, and the populist age according to my account of the CBC's history. Its death must be our hope, as well as our despair.

13 | On Standing Back to See Better

> Always assume that there is one silent student in your class who is by far superior to you in head and heart.
>
> —*Leo Strauss, when asked about his approach to teaching*[1]

> No matter what particularities … we bring to bear on public discourse, the moment of apprehending something as public is one in which we imagine, if imperfectly, indifference to those particularities, to ourselves.
>
> —*Michael Warner*[2]

When I began broadcasting on CBC Radio in 1971, I did so as an enthusiastic acolyte of Brazilian educator Paolo Freire. In his then-recently published *Pedagogy of the Oppressed*, the book that had inspired me, Freire had declared, without qualification, that "the correct method [of education] lies in dialogue."[3] Dialogue was the defining feature of "true praxis," the heart of authentic communication, and the "essence of revolutionary action."[4] Anything else, Freire insisted, was mere "banking," in which "one person deposits ideas in another."[5] I was captivated, as were many of my contemporaries. The romance of dialogue was in the air. (Another of many possible examples from that time was the popularity and pervasive influence of Martin Buber's *I and Thou*.) But I was also perplexed, since the field I was entering appeared to be, inescapably, a monologue. This was in the infancy of the answering machine, and before CBC Radio programs began to invite "feedback" from their listeners, but, even with today's tools for instant response, broadcasting remains, structurally speaking, a one-way communication. Reaction must always come after the fact. So how, I wondered, can broadcasting retain something of the character of a dialogue, even as it addresses an audience that is for the most part unseen, unknown, and unresponsive? The question has stayed with me ever since.

My attempts to answer this question over the years have revolved around the idea that the absent listener can be made an interlocutor only when the broadcaster's way of speaking makes room for such a response. This hospitality, or making room, can only be accomplished, I have concluded, by the cultivation of a certain formal distance between broadcaster and listener. In what follows, I will explain what I mean, but the first thing to be said—obvious enough, perhaps—is that my conclusion runs very much against the grain of the era in which I worked at the CBC—the populist era, as I have called it. The trend during this period has been toward broadcast talk that is casual, familiar, and obliging: one has wanted to get closer to listeners, not farther away. Listener, or viewer, and broadcaster are assumed to be people who know the same things, want the same things, and believe the same things. The broadcaster speaks as to a close friend with whom much can be taken for granted. ("Hey there, Jessie," was how I recently heard a guest host on CBC Radio's Toronto afternoon show begin an interview with a university teacher from Montreal, whom he was apparently meeting for the first time.) Formality tends to be seen as an unwelcome reversion to the stuffiness that was supposedly characteristic of the CBC's first era. On CBC Radio, even documentaries are now usually presented in the offhand, confiding, by-the-way tone popularized by American public radio programs like *Radio Lab* and *This American Life*—both currently carried on CBC Radio. In television documentaries, formal narration and exposition are usually avoided altogether in favour of the illusion that the program has, in effect, assembled itself. Formality and distance are out.

So why should I now hope to get a hearing for my against-the-grain conclusion? One reason is that I continue to think it has merit: Leaving space for the listener to think still seems to me the only way in which an element of reciprocity can be incorporated into broadcasting's centrally produced monologue. But, more than that, I think that the times have changed—in two crucial ways. First, as already discussed, the *consensus* assumed by the CBC's populist, *Canada Lives Here* manner has turned into profound *dissensus*. Canadians now disagree fundamentally and, for the time being at least, irreconcilably; and, insofar as the CBC tries to maintain the fiction of *consensus*, it disqualifies, disparages, and antagonizes the divergent publics who don't belong to, or identify with the CBC's confident but outdated *we*. Second, we now face the increasingly acute problem of media in general. Media have directed people's attention since the Ten Commandments and the Code of Hammurabi were written on stone, but they have never before achieved the total enclosure of attention that the contemporary acceleration seems to threaten. Both these predicaments—dissensus and hyper-mediation—seem to me to urge a questioning of populism's founding precept—identify with the audience and encourage the audience to identify with you.

The CBC needs to step back and consider its position. A formal space, or forum, able to accommodate currently antagonistic standpoints needs to be constructed. Along with it, a formal manner appropriate to the curation of such a space needs to evolve—a manner that I imagine as something quite new, but also as referring to and recuperating older styles. Finally, mediation itself has to be put into question in the face of what otherwise threatens to become terminal disorientation. This too will require a recovery of formal distance between the public broadcaster and its audiences. Mediation can only be counteracted and limited when it has first been brought to light. The populist project of trying to make intermediation disappear must now give way to something like its opposite—an attempt to replace the shrunken space and foreshortened time of identification with an expanded critical distance.

The dilemma I want to address has been building up to its present intensity for a long time. Ever since the invention of the telegraph some two hundred years ago, communications media have been gradually and progressively collapsing distances. By the 1850s, American writer Henry David Thoreau was already lamenting his compatriots' habit of treating the far away as if it were the near at hand. "Hardly a man takes a half-hour's nap after dinner," he wrote in *Walden*, "but when he wakes. he holds up his head and asks, 'What's the news?'"[6] The gramophone and the telephone, the radio and the television all intensified this tendency and increased the intermingling of the near and the far. The uncanny voice from everywhere and nowhere was domesticated. Movie stars, beloved singers, and broadcast "personalities" became mediators and intercessors—figures from an ideal realm who offered solace, understanding, and communion. At first, mass media observed a certain decorum, behaving, for the most part, as courteous guests who recognized the threshold they were crossing in entering people's homes and lives. The addressee of mass media retained, for a while, the status of an honoured stranger.

This has changed gradually but steadily throughout the populist era, to the point where it now makes sense to speak, as philosopher Jean Baudrillard does, of an implosion—an inward collapse in which old distinctions between private and public, as well as between reality and representation, have tended to disappear. The capabilities of new media—constant surveillance, constant exposure—have a lot to do with this, but it is also a goal that has been actively pursued under the banner "the personal is political."[7] The idea has been that oppression and domination flourish wherever publicity is withheld, so everything formerly considered private ought to be made public. What has followed might be called a crisis of indistinction, or, alternately, disorientation.

When the private is public, the personal is political, and the self is turned inside out, in the way allowed by social media, it is difficult for people to know where they are, what they are, or whom they might be addressing. Where each one makes their own sovereign declaration of identity, the public as a place of interchange tends to dissolve.

The Fall of Public Man

One of my first hints that the character of the public might be threatened by this new emphasis on authenticity and identity, self-discovery, and self-disclosure, was Richard Sennett's prescient book, *The Fall of Public Man*, which appeared in 1974. Like me, Sennett had been taken with the personal-is-political ethos of the 1960s. But then he had begun to fear that the taste for self-authentication and "getting real" was beginning to produce what he called a "city of ghettoes"—a world of non-communicating solitudes.[8] Knowing oneself, he said, had become "an end in itself, not a way to know the world."[9] The "boundaryless self"—Sennett quotes the phrase from American literary critic Lionel Trilling—was erasing the very condition on which public life depends—a clear distinction between the public and the private.

Sennett set out to define this dialectic and to vindicate discredited modes of public interchange—artifice and convention, theatricality and playfulness. His central argument was that, if there is to be a common world, it must be a world at a certain remove from the urgencies of private identity and private concern. Artifice, he said, enables sociability by holding the real in suspension—not denying it but taking a distance from it in order to create a common ground on which it can be investigated without overwhelming the investigators. The public and the private ought, in this way, to limit and define, restrain and glorify each other—spontaneity abetted by convention, private feeling intensified by public restraint.

One of the ways in which Sennett illustrated his conception of the proper character of public life was by the image of play. Drawing on Dutch historian and philosopher Johan Huizinga's *Homo Ludens*, he pointed to a number of features of play which make it a model of public interaction. Play is "disinterested" in the sense that the activity is its own reward—one plays in order to play without regard for the outcome. Play is purely voluntary—no one has to do it. Play is secluded—it occurs in some dedicated setting that has been created or adapted for that purpose. And play takes place "at a distance from the immediate desires of the self"—an effect partly achieved through a set of impartial rules. In play, convention promotes freedom and spontaneity, and the ability to play encourages the idea, as Sennett says, that "worldly conditions are plastic"—they can be reinvented, though only within the agreed and understood parameters that enable common action.

Sennett's *The Fall of Public Man* encouraged and supported the direction my own thinking was beginning to take at the time I discovered his book. I had concluded that the public had to be a space in which every opinion—excepting those that threaten the space itself—was able to find a place and a voice. But, from my vantage point at CBC Radio, it was difficult to see how this could happen in a radio service that was increasingly stylized and increasingly dominated by populist we-making.[10] Sennett gave me a crucial clue. His concern was public life in general, with only passing attention to media, but he was drawing attention to something that addressed the question I was asking quite directly. If people who are in some way different from one another are to be able to interact with one another in public, there must a space in which this can happen—a space created by courtesy, restraint, and the various other formal codes that construct the public realm. It should be the same, I thought, in broadcasting. A public forum open to all must be a formally constituted space in which distances and differences are respected, and even celebrated, rather than being collapsed, foreclosed, or otherwise explained away.

In this view, distance and formality *enable* freedom and spontaneity rather than opposing them. Even Hannah Arendt, a political thinker who defined the public in strictly impersonal terms, still held that self-disclosure—the revelation in act and speech of "who I am"—was the very essence of public life. Sennett was less stringent than Arendt—allowing more of culture and sentiment into his account of the public—but he pointed to the same paradox: We develop understanding and sympathy for one another when there is a sufficient interval or intermission between us to allow these dispositions room to grow and take root. We get closer by standing back. To acknowledge this paradox is not to speak for formality as a good in itself, or discount the role of more intimate knowledge in generating sympathy, but only to recognize that there must be a space across which sympathy can spark. Intimacy may follow. But where intimacy is assumed, sympathy compelled, and identity imposed, what Arendt calls "the space of appearance" collapses.[11] Brought too close too soon, people are driven apart. The violent resentments currently agitating enemy publics are the result.

What I mean by too close and too soon is not easy to say. The habit of disregarding distance, foreshortening time, and short-circuiting any formal obstacle to easy-going intimacy has become so ingrained that it may seem to many that nothing can come too close or occur too soon. Who can discern the right time when all the world's temporalities are scrambled and super-imposed in the simultaneity of McLuhan's global village? Who dares to speak for distance at a time when injury and insult, reparation and apology dominate political agendas, and excited sympathy is the very *raison d'être* of public life? It might seem perverse in such a climate to suggest that the CBC ought to begin to try to keep its distance from its audiences. Yet it appears to me that only formal distance, and a formal sense of occasion, stand any chance of rehabilitating the CBC as a public forum in which the country might air its differences and thus begin to get to grips with them.

Critical Distance

Distance is, first of all, critical distance—the remove at which it becomes possible to understand something on its own terms. By *understand*, I don't mean to suggest acceptance, agreement, or assent but only some intuition of how and why a given object is as it is. Nor do I mean to suggest by *on its own terms* that anything can ever be apprehended entirely in this way—its terms will always have to be adjusted to my terms according to the principle that Hans Georg Gadamer calls "fusion of horizons." But, that said, I do think it is possible to bracket one's own assumptions sufficiently to frame a question that begins, at least, to disclose and expand the scope of what is to be understood. And, such a question having been asked, there must then be a space—a clearing, let's say—in which the answer can unfold. The question cannot contain or imply its answer, and so it must come from a distance that leaves leeway for surprise, contradiction, or complication. Only at a distance can I put the prejudice by which my question will inevitably be coloured into play, and only from a distance can the other find the opening their answer will require. Such distance is neither remoteness nor indifference, but is rather an allowing, a waiting, and a hoping that withdraws or withholds itself in order to leave room for the other.

I do not often come across this disposition in today's CBC. The CBC is pervaded, rather, by a spirit of positivity—it wants to saturate the space between itself and its audiences with everything that is pleasing, therapeutic, and uplifting. Questions without prescribed answers are avoided. The consensus of its audiences on all important questions is taken for granted. Disagreement on any crucial point is generally seen as evidence of either bad faith or misinformation.

Acclamation rather than inquiry is the dominant note. What lies in between the broadcaster and the listener is not brought into question. By in between, I mean to indicate not only physical media—the wires and waves, satellites and studios on which broadcasting depends—but also what might be called rhetorical media—the conversational conventions by which the CBC constructs its relationship with its audiences. Media, as I've stressed, are inherently ambiguous. Like Simon Weil's image of the two prisoners who are both separated and joined by the wall between them, media comprise both flow and resistance, facility and obstacle, ability and disability. Populism has promoted identification. "That's what I think," Douglas Leiterman hoped *Seven Days'* viewers would exclaim. Margaret Lyons wanted listeners to recognize "their insights" and their "deeply felt yearnings" in the CBC's obliging mirror. Differences, distances, and devices that might interfere with this desired identification were downplayed. The contradictory, two-faced character of media was largely forgotten. "Ain't nobody here but us chickens" was the prevailing attitude. Peter Gzowski really *was* the bosom friend that so many said they felt they were losing when he retired, and not just a skillful rhetorician, whose carefully cultivated art was to sound artless.

I'm not depreciating this art—only asking that it be recognized and analyzed. Peter Gzowski gave the populist ethos of the CBC's second era one of its consummate expressions. For those who wanted to belong to his club, he made differences and distances disappear. But now a confused and divided country needs the CBC to perform a new part—more counsellor than comforter, more gadfly than reassuring friend, more convenor and curator than confidant.

I do not pretend to know how this new style might look and sound. It will have to be invented in practice, in the same way that Gzowski, and hundreds of others, invented the populist style. What will have to come first is conscious awareness of the mediations and rhetorics, distances and devices that the prevailing style has tended to overlook and even dissimulate in its rush to gain the public's confidence. Populism has wanted to serve and to satisfy its audiences—to reach them where they are, as I often heard during my years at CBC Radio. The "program guidelines" that I analyzed in my chapter on Creative Renewal specified the "strategically correct" manner by which we could gain and hold the attention of our "target audience." I am talking, rather, of a style that would treat the public as undetermined and focus on making publics rather than anticipating them. I recall, in this connection, a discussion I once had with my old friend and *Ideas* colleague Max Allen—a conversation we both still sometimes recollect as a touchstone in our decades-old dialogue about public broadcasting. He had asked me about the intended audience for a series of mine. I found myself quite unable to answer. My inability made me realize not only that I had no specifiable class of listeners in mind, but also that I didn't want to imagine any audience in advance of my addressing it. I wanted listeners, certainly, and I wanted to make my subject as transparent and as vivid as I possibly could. But anticipating who would listen and respond seemed to me to foreclose and pre-empt the distance about which I've been trying to write here. I hoped to assemble a public, not gratify a clientele. And, to that end, the work should somehow remain open and unfinished—in search of its listeners rather than able to describe them in advance. These listeners always turned up—though sometimes long after the fact and not always as the people I expected—but I could not possibly have aimed at them without denying them the space to think with me that I was trying to preserve and, in part, create.

The manifesto that began the first broadcast of *This Hour Has Seven Days* expressed the program's intention to become "mandatory viewing." This emphasis on irresistibility was sustained through many subsequent incarnations and iterations of the populist project of predicting and prescribing, surrounding and saturating its audiences. The aim was to be compelling in all of that word's several senses. One casualty, as I've said, was distance—that distance at which the audience can remain aware of the various intermediations that separate it from a given broadcast source. Another was the authority of subjects which tended to diminish as the exclusive authority of the audience increased. My claim is relative, of course. No program can afford to completely ignore the conventions of the milieu on which it depends. Authors wanted to take part in CBC Radio's long-running *Writers and Company* because Eleanor Wachtel was an alert interviewer and a sensitive and educated reader of their books. The even longer-lived *Quirks and Quarks* has kept the trust of the scientists on whom it depends by its knowledgeable respect for their sciences. Some music programs, like *Saturday Night Jazz* or *Saturday Night Blues*, remain steeped in the standards and traditions of their respective genres. But the trend nevertheless has been away from the idea that broadcasters owe an obligation to their subjects which is just as weighty as their obligation to their listeners. When a producer from CBC Music turns up on my local afternoon show to introduce "songs you need to hear"—a recurring item—I don't have the impression that any critical claim whatever is being made for the music. This is music that *I need to hear* because it has somehow achieved a transitory cultural fit—it's created a ripple in the "the culture," or the "media ecosystem," or some other indistinct organ of self-replicating consensus. No one is claiming that the songs are good, according to some articulate or enduring standard. Music—as a discipline, as a set of traditions, as a body of taste, as, in short, a subject—is beside the point.

Another example that recently struck me of the diminished authority of subjects occurred during an *Ideas* broadcast entitled "Bring Back Grumpy George: The Forgotten Message of George Grant." George Grant, in his time

(1918–1988), was a landmark in Canadian intellectual life, and a major contributor to the CBC as well.[12] But when Grant was revived on *Ideas*, he was not treated with the deference and circumspection which his work in philosophy and his place in Canadian life might seem to have earned him. Instead, it seemed that Grant was auditioning. "What's [he] going to do for us?" asked one of the show's two presenters. The other wondered whether elements in Grant's thought that she found unfashionable should be "given a pass." They speculated about whether Grant might not currently be "a hard sell."[13] My strong sense was that Grant was required to submit to their judgment, but they were not required to submit to his. What matters is what's happening *now* with *us*. And, to that end, historical characters like "grumpy George" must bow and present their credentials. What Harold Innis called "the elements of permanence essential to cultur[e]" are washed out.

Rhetoric Reconsidered

Earlier in this chapter, I used the word *rhetoric* to try to convey the ways in which the space between the CBC and its audiences is envisioned, embodied, and brought to life. The term long ago went out of fashion and began to ring of inauthenticity, but its history illuminates so exactly the difficulty in which the CBC now stands that I think this history is worth reviewing. Once upon a time, rhetoric simply denoted the art of persuasion, or of public speech, since the two were assumed to be the same. The study of rhetoric was considered one of the foundations of the liberal arts.[14] Today, the word rhetoric generally refers to something artificial and insincere—a rhetorical question is asked purely for effect, political rhetoric implies dissimulation of some kind. Ever since the Royal Society swore to speak and write with "mathematical plainness," depreciation and disavowal of rhetoric have been hallmarks of the modern age.[15] Rhetoric, however, did not disappear. It would be truer to say that it was repressed or forced underground. A historian of the subject, Walter Ong, says that in the modern period, under the guise of elimination, rhetoric was actually "disrupted, displaced, and rearranged."[16] It became a "bad" word. Proponents of science contrasted its "softness" with the "hardness" of knowledge gained by experiment or calculation. Romanticism condemned rhetoric's "contrivance" in favour of freer and more spontaneous expressions that bubbled up "unsolicited" from the "inner wells of being."

In modern times, the taste for authenticity and the taste for objectivity have flourished together. They are opposing ideals, certainly, but they are also complementary parts of the modern stance, each compensating for the other. Objectivity has stood for pure representation—truthful, accurate, disinterested; authenticity for pure expression—the voicing of the original and the identical, the announcement of who I am. Both stances overlook the media on which they depend. But when rhetoric is put out of sight, out of mind, and out of account, in this way, it potentially becomes a lot more powerful than when it was acknowledged, cultivated, and studied. To take the example of broadcasting: When broadcasters become unable to recognize the rhetorical character of their utterances, they lose sight of the interested, persuasive character of their speech. They begin to take their own good will and good intentions for granted. Patrick Watson and Douglas Leiterman, at the inception of the populist revolution, could assume that a CBC broadcast was a proper vehicle to "right wrongs," without having to consider the consequences of making television the vehicle of justice, and themselves its agents. They thought of themselves not as interested parties, who owed their publics an account of these interests, but as authentic voices welling up out of an abundance of care and concern.

Identification with the audience, and the audience's responding identification, once made this stance convincing. Think of the passion stirred by *This Hour Has Seven Days* or the love listeners felt for Barbara Frum or Peter Gzowski. But identity has now become a dead-end—acting both as a fetter on thought and a provocation to the many citizens not included in its tight embrace. This certainly doesn't counsel a revival of rhetoric, in the pejorative sense, but it does urge a recognition of the motivated and interested character of broadcast talk. Only through such self-awareness will the CBC become capable of enlarging the public square and augmenting understanding within it.

The Character of Public Life

Theorists of the public, from Jurgen Habermas to Hannah Arendt, have defined the public sphere as the arena of our most real and most vivid existence. For Habermas, it is the site of that free interchange which leads to the formation of a "public opinion" strong enough, as Habermas says, "[to] compel public authority to legitimate itself before [this] opinion."[17] For Arendt, it is the space of "appearance"—of things "seen and heard by others as well as ourselves"—in which reality itself is constituted.[18] Life in public, she says, embodies "the highest possibility of human existence," and "to be deprived of it is to be deprived of reality."[19] But what these writers describe is more or less the direct opposite of what people currently experience. It is in private that we feel ourselves to be most real and most alive. This was what American sociologist Nina Eliasoph discovered when she tried to put the claims of theorists like Arendt to the test. The farther "backstage," or out of the public, her research subjects were, she reported, the freer they felt and the more frankly they spoke. Most considered the public the realm of self-interest and not of free interchange, and this "folk definition of the public sphere," Eliasoph wrote, "kept most interesting debate out of public circulation."[20] Her conclusion rings true, but demands explanation.

Historically, public and private have been reciprocally defining ideas complementary parts of a single regime, each defining, enabling, and extending the other. In the eighteenth century, public opinion gestated in newly secured private spaces, like homes and salons; new areas of privacy were carved out by public action to restrict the power of the state. But, during our time, and more gradually during the whole industrial age, the necessary balance between these mutually opposing and mutually sustaining realms has been disturbed. The private has invaded the public, and the public the private, to the point where there is effectively no boundary between them. Privacies of every kind flood the public square; publicity emanates from the former strongholds of privacy. This, I think, is why Eliasoph is reduced to paradoxes in speaking of how "public spirit" is expressed in private but "evaporates" in public.[21] The words retain a rudimentary, common sense meaning—one can still be in private or in public—but the distance from the public that is proper to privacy is lost, as is the distance from the private that defines a distinctly public sphere. Both are deprived of their grammar and their structuring conventions. Privacy provides little shelter or seclusion; the public become a site at which preformed brands, identities, and ideologies are projected before a cynical, disbelieving, and increasingly angry audience. We suffer what I earlier called a crisis of indistinction—no one knows quite where they are, or what time they are living in; no one is sure how to speak or act. The only remedy, and the task I am enjoining on Canada's *public* broadcaster, is a rediscovery, reappropriation, and renewal of the public as a distinct and formally constituted sphere.

If Canadians are to have a common life, they must first have a common world. And, if there is to be such a world, it will first have to be made, because the many worlds we currently inhabit are manifestly at odds—incoherent within themselves, as much as amongst themselves. Such world-making is a public undertaking, but it cannot begin until we have gained—and, to an extent, regained—an understanding of the nature of the public. The public cannot continue to be what it now mainly is—a sphere in which private identities jostle for attention, private agonies seek iconicity, and private meanings compete for hegemony. It must become a place, speaking metaphorically, where meanings are open to investigation, and the business of world-making is made conscious. All the ways in which the world is made, as it were, behind our backs must be challenged—from the immaculate conception of the news, to the hundreds of other crises, emergencies, and issues that unseen stage machinery thrusts readymade on our attention. For this we will have to go beyond ourselves, beyond our private persons. This is the character of the public that I would like to emphasize—that it is a sphere in which we are not quite ourselves, but take a distance from ourselves in order to find a ground on which to meet others.

American writer Michael Warner, in his book *Public and Counterpublics*, speaks of this quality of being, at once, oneself and not oneself as a "double voicing," and makes it a crucial condition of public life.[22] To speak with strangers, one must venture out of identity and onto new ground, trying on idioms and manners not initially one's own—playing a part. It is what is particular about us—some particular interest or attribute, advantage, or disadvantage—that brings us into the public in the first place, but once there we can only find agreement with others by also bracketing these particularities. "The moment of apprehending something as public," Warner writes, is a moment at which we must try to imagine, however imperfectly, "indifference to those particularities, to ourselves."[23] This indifference to oneself was what was novel about the bourgeois public sphere that took shape in the eighteenth century. This new arena embodied what Warner calls "a principle of negativity"—the validity of what was said in public would derive from its inherent merit and not from who said it. It would "carry force not because of who you are but despite who you are."[24] The ideal of disinterest developed—"the public opinion" was considered to be an opinion arrived at by setting aside self-interest. People cultivated a capacity for abstraction—a capacity to stand back, or withdraw, from own's own particular position and personality, just as the roots of the word abstract say.[25]

Warner admits that the idea of the public as a space of doubleness and disinterest was, from the outset, "utopian." It ran against the strong preference that he finds in Western thought for identity, consistency, and non-contradiction. "From Rousseau to Reagan," he writes, "self-unity has been held to be a public value."[26] And it was also capable of being made a source of domination because at first the stance of "self-abstracting disinterestedness" was open only to literate, propertied, white males. Only their persons were normal, and therefore unobtrusive. Their identities were, as grammarians say, unmarked.[27] They could free themselves from the toils of identity because they could take their own identity for granted. *Disinterest* was, so to speak, their privilege. Others—the majority of women, for example—would not so easily have been able to hold their identity in abeyance, without at the same time consenting to give up their vital difference from the governing norm.

This is a crucial point in Michael Warner's book and accords with what philosopher Charles Taylor also says in his influential essay, "The Politics of Recognition."[28] The tension between identities-demanding-recognition, and the indifference-to-identity by which, as Warner says, one "apprehends something as public," must somehow be maintained so that neither trumps the other. Michael Warner's own thinking took shape in the context of the gay struggle against the monopoly of heterosexual norms, and he has always insisted that understanding of the public must have room for the very particular kinds of publics to which queer culture and queer demands for recognition give rise. Yet, he still argues against pure identity of the here-I-am-like-it-or-lump-it variety. In becoming "public subjects," he says, we must necessarily become "non-identical with ourselves."[29]

There should be a balance then between the legitimate demands of identity and the self-distancing which is an indispensable condition of a genuinely public life. But this balance has been lost. From disinterestedness as the unmarked privilege of propertied white males, we have swung to the opposite extreme. Identity is now the unmarked privilege of whomever can claim historical disadvantage. One can speak only "as a woman" or "as an Indigenous person" or even, with the requisite apologies, "as a white male." In the name of counteracting privilege, a new privilege has been created, but one that is fatal to the public. The move that was undertaken to save the public square from the dominating fiction of disinterest, and the mid-Atlantic accent, has ended up destroying its very basis. What is now taking place looks to me increasingly like a war in which sovereign and extremely well-fortified identities compete for advantage. In such a milieu, every gesture, by definition, either reinforces or threatens identity. Privilege, once exercised, becomes an ineradicable stain. And, since all's fair in war, identity increasingly becomes a weapon to be deployed opportunistically—cast off as an encumbrance when claiming equal right, and put back on when claiming privilege, redress, or some other advantage. No one can afford to be non-identical with themselves, and so public conversation dies. "The very mechanism designed to end domination," Warner says, becomes itself "a form of domination."[30]

No one can know, at this point, whether a new public forum is possible, and, if so, what form it will take. Historical images of the public sphere are equally uncertain. Lively academic disputes continue about whether the public sphere described by thinkers like Habermas and Arendt ever really existed in quite the form in which those thinkers imagined it. All one can do at this point is to try to identify the conditions that currently prevent such a forum from beginning to sprout and grow. A few things, however, are clear. Without a place of free exchange, Canada will remain locked in obscure and unavailing ideological combat. Without unconstrained and unpoliced freedom of speech and thought, such a place will never be constructed. And without a new mode of address which tries to create space rather than saturating it, the CBC will never regain the ear of that part of the country whose attention they have lost. A clearing will have to open, and the CBC will have to learn, once again, how to stand back and allow space and time for conversation.

Notes

Introduction

1. David Cayley, *Turning Points in Public Broadcasting: The CBC at 50*, *Ideas*, CBC Radio, November 3, 10, 17, 24, December 1, 1986, transcript pp. 28–29, https://www.davidcayley.com/transcripts. The creation of this *Ideas* series gave me the opportunity to interview many of the CBC's founders, and I will refer to it frequently throughout the first half of the book, hereafter as just *Turning Points*.
2. Eric Koch, *Inside Seven Days: The Show That Shook the Nation*, Prentice Hall, 1986, p. 216.
3. Ibid., p. 136.
4. Ibid., p. 250.
5. Ibid., p. 12.
6. *Turning Points*, p. 15.
7. Ibid., p. 30.
8. *Inside Seven Days*, p. 249.
9. Richard Stursberg, *The Tower of Babble: Sins, Secrets and Successes Inside the CBC*, Douglas and McIntyre, 2012, p. 11.
10. Ibid., p. 310.
11. Patrick Watson, *This Hour Has Seven Decades*, Toronto McArthur and Co., 2004, p. 144; Ross McLean was the executive producer of *Close-Up*, the television newsmagazine where Patrick Watson and Doug Leiterman, the other creator of *Seven Days*, cut their teeth as television producers.
12. https://www.youtube.com/watch?v=5I0tk6OO5sw; https://globalnews.ca/video/8542159/trudeau-says-fringe-minority-in-trucker-convoy-with-unacceptable-views-dont-represent-canadians.
13. *Let Us Compare Mythologies* (Contact Press, 1956) was Leonard Cohen's first published poetry collection.
14. Frank W. Peers, *The Politics of Canadian Broadcasting 1920–1951*, University of Toronto Press, 1969, p. 101.
15. *Passion and Conviction: The Letters of Graham Spry*, ed. Rose Potvin, Canadian Plains Research Center, University of Regina, 1992, p. 38.

16. https://www.poetryfoundation.org/poetrymagazine/poems/26063/hypocrite-auteur.
17. Octavio Paz, "Reflections: Mexico and the United States," *The New Yorker*, September 17, 1979, p. 153.
18. Mark 2:22.
19. Graham Spry, "The Decline and Fall of Canadian Broadcasting," *Queen's Quarterly*, Summer, 1961.
20. Wayne Skene, *Fade to Black: A Requiem for the CBC*, Douglas & McIntyre, 1993.
21. David Taras and Christopher Waddell, *The End of the CBC?* University of Toronto Press, 2020.
22. https://cpcassets.conservative.ca/wp-content/uploads/2023/11/23175001/990863517f7a575.pdf, p. 35.
23. *Analysis of Government Support for Public Broadcasting and Other Culture in Canada*, report done for the CBC by consultant Nordicity in April of 2011, p. 4.
24. See Chapter One, p. 19.

Chapter 1

1. David Cayley, *Turning Points in Public Broadcasting*, Canadian Broadcasting Corporation, 1986, p. 1. This is the series, mentioned in the Introduction, that I prepared for *Ideas* in 1986, the fiftieth anniversary of the creation of the CBC. Audio of the series is here: https://www.davidcayley.com/podcasts/category/CBC. The transcript, to which I will refer, is here: https://www.davidcayley.com/transcripts. I have made extensive use of this series as a source for this chapter. Whatever historical material is not footnoted will be found there.
2. https://newsm.org/people/fessenden/.
3. https://montrealgazette.com/entertainment/local-arts/the-station-that-would-become-cfcf-made-radio-history-100-years-ago-in-griffintown.
4. *Turning Points*, p. 2.
5. *Turning Points*, p. 3.
6. http://news.bbc.co.uk/aboutbbcnews/spl/hi/history/noflash/html/1930s.stm.
7. *Turning Points*, p. 4.
8. *Passion and Conviction:* p. 38; see also Bruce Steele's film documentary about Spry—https://vimeo.com/506303879.
9. *Passion and Conviction*, p. 78.
10. *Turning Points*, p. 4.
11. Ibid, p. 5.
12. "The New Canada Movement," *The Farmer*, December 1933, https://www.farms.com/reflections-on-farm-and-food-history/historical-articles-archive/the-new-canada-movement.
13. *Turning Points*, p. 6.
14. Ibid, p. 7.
15. Ibid, p. 8.
16. Ibid., p. 8.
17. Ibid, p. 9.

18. A good source on Alan Plaunt, and his dispute with Murray, is Michael Nolan's *Foundations: Alan Plaunt and the Early Days of the CBC*, House of Anansi, 1986.
19. *Turning Points*, p. 11.
20. Ibid., p. 11.
21. Ibid., p. 11.
22. Ibid., p. 12.
23. The incident receives only a passing mention in Frank Peers, *The Public Eye: Television and the Politics of Canadian Broadcasting 1952–1968*, but it is dealt with a length by historian Christopher Dummitt in an article called, "Harry Ferns, Bernard Ostry, and *The Age of Mackenzie King*: Liberal Orthodoxy and its Discontents in the 1950s." (*Labour/LeTravail*, 66 (Fall 2010), 107–139). In what follows I rely on Dummitt's article.
24. *Turning Points*, pp. 12–13.
25. https://www.thecanadianencyclopedia.ca/en/article/canadian-broadcasting-corporation.
26. Christopher Plummer, interviewed by Eleanor Wachtel on CBC Radio's *Writers and Company*, July 2, 2021—this was when I heard it; the interview was recorded and first broadcast in 2011.
27. *Turning Points*, p. 17.
28. *The Politics of Canadian Broadcasting 1920–1951*, p. 187.
29. A full account of this episode is given in *The Politics of Canadian Broadcasting, 1920–1951*, pp. 378–382.
30. Political scientist Gad Horowitz coined this term in his 1968 book *Canadian Labour in Politics* (U of T Press), but some of the people to whom he applied it, like philosopher George Grant, deprecated this label. (See my *George Grant in Conversation*, House of Anansi, 1995, p. 104.)
31. Herschel Hardin, *A Nation Unaware: The Canadian Economic Culture*, J. J. Douglas, 1974.
32. *The Politics of Canadian Broadcasting 1920–1951*, p. 341.
33. Ibid., p. 363.
34. *The Public Eye: Television and the Politics of Public Broadcasting, 1952–1968*, pp. 160–161.
35. *The Politics of Canadian Broadcasting 1920–1951*, p. 118.
36. *Turning Points*, p. 40.
37. Ibid, p. 40.
38. Ibid., p. 41.

Chapter 2

1. *The Public Eye*, p. 11.
2. Ibid., p. 24.
3. Knowlton Nash, *Cue the Elephant: Backstage Tales at the CBC*, McClelland and Stewart, 1996, p. 84.
4. Graham Murdock, "Radical Drama, Radial Theatre," *Media, Culture and Society*, 2: 2, April, 12980, pp. 151–168.

5. *Turning Points*, p. 18.
6. "The Gentlemen's Strike" was the title Barbara Fairbairn gave to her MA Thesis for Carleton University on the strike and its consequences. See: https://curve.carleton.ca/451600a1-d551-41a3-8fea-64cac79a3db0.
7. *Turning Points*, p. 24.
8. *The Politics of Canadian Broadcasting, 1920–1951*, p. 340.
9. *This Hour Has Seven Decades*, p. 109.
10. *This Hour Has Seven Decades*, p. 123.
11. *Turning Points*, p. 28.
12. Ibid, p. 28.
13. Ibid., pp. 28–29.
14. Ibid., p. 30.
15. *Turning Points*, p. 31.
16. *This Hour Has Seven Decades*, p. 179.
17. Ibid., p. 248.
18. *Inside Seven Days*, p. 254.
19. Ibid., p. 136.
20. Ibid., p. 216.
21. Ibid., p. 216.
22. The AVRO Arrow was a delta-winged military aircraft, created by the A. V. Roe Company in Malton, Ontario, during the 1950s. It had superior capabilities to other military aircraft of the time and what many thought was a visionary design, but its development was halted by the Diefenbaker government in 1959 and its prototypes were destroyed.
23. https://www.cbc.ca/player/play/1403659265.
24. T. S. Eliot, "East Coker," *Four Quartets*, Harcourt Brace, 1943, p. 20.
25. I owe the word *ecstatic* to media critic Mark Crispin Miller in his *Boxed In: The Culture of TV* (Northwestern, 1988, p. 16), though Miller speaks of McLuhan's "mock-ecstatic phase."
26. "The Politics of Information," *Ideas*, May 1, 1983, p. 8, https://www.davidcayley.com/transcripts.
27. The Supreme Court judgment is here: https://scc-csc.lexum.com/scc-csc/scc-csc/en/item/4045/index.do?site_preference=normal; a transcript of the *Seven Days* item is here: https://historyexhibit.waypointcentre.ca/exhibits/show/legislation/item/229.
28. Junius was the pseudonym of an unidentified writer who contributed a series of letters to the English newspaper the *Public Advertiser between* 1769 and 1772.
29. The question, often quoted in Latin as *Quis custodiet ipsos custodes,* traces back to the *Satires* of Juvenal, written around 100 CE.
30. It was actually his translator, Thomas Burger, that bestowed this name when he rendered Habermas's *Strukturwandel der Öffentlichkeit* as *The Structural Transformation of the Public Sphere* (MIT, 1989). Habermas's German word *Öffentlichkeit*, literally publicness, or publicity, has quite a different valence than does public sphere.
31. Oscar Wilde, *The Soul of Man Under Socialism*, Max N. Maisel, 1915, p. 40.

32. "*The Legacy of Harold Innis*," *Ideas*, CBC Radio, December 13, 1994, transcript p. 16; the transcript is here: http://www.davidcayley.com/transcripts.
33. There was a young *lady* of Niger/ Who smiled as she rode on a *tiger*;/ They returned from the ride/ With the *lady inside*,/ And the smile on the face of the *tiger*.
34. Alex Barris, *The Pierce-Arrow Showroom Is Leaking: An Insider's View of the CBC*, Ryerson, 1969, p. 95; "the mandate" is CBC shorthand for the corporation's statutory purposes.
35. *The Tower of Babble: Sins, Secrets and Successes Inside the CBC*, p. 11.
36. *The Way of Life According to Lao Tzu*, trans. Witter Bynner, Capricorn, 1962, Stanza 27, pp. 34–35.
37. *This Hour Has Seven Days*, March 20, 1966; Ouimet's response is here: https://www.cbc.ca/archives/entry/1966 hosts on air tear hastens end of-seven-days.
38. Peter Hughes, "The Eighth Day of the Week," *Canadian Forum*, January, 1967.
39. *Turning Points*, p. 13.
40. *The Tower of Babble*, p. 310.
41. The phrase "the bubble reputation" comes from Jacques' "Seven Ages of Man" speech in Shakespeare's *As You Like It*, Act Two, Scene Seven.
42. I was introduced to this idea by Ivan Illich during a recorded conversation which is reproduced in David Cayley, *Ivan Illich in Conversation*, House of Anansi, 1992, p. 65 ff.
43. William James, *Principles of Psychology*, Henry Holt, 1918, p. 488.
44. George W. S. Trow, *Within the Context of No Context*, Little, Brown and Company, 1981.

Chapter 3

1. The committee was chaired by Robert Fowler, who had headed the Royal Commission on Broadcasting that reported in 1957. The CBC's report was prepared and presented by Vice-President Eugene Hallman, who appeared in the previous chapter as the man who laid down the law to the producers of *This Hour Has Seven Days* in 1964.
2. Clay Carter, "The CBC Radio Revolution 1964–1976, A Re-examination," p. 112. This is an MA thesis, for Carleton's School of Journalism, available here: https://curve.carleton.ca/8ff87603-cf85-4196-befb-6703294df705. Carter's thesis, as far as I know, is the only history that has been written of the radio revolution, excepting personal memoirs, like Barbara Frum's, cited below.
3. Ibid, p. 18.
4. Sandra Peredo, "Can Radio Make a Comeback?", *Maclean's*, December 1, 1968.
5. Karin Wells' documentary, "A Tribute to Margaret Lyons" was released on January 28, 1921, and can be found here: https://www.youtube.com/watch?v=elkrJJUw3LY. Subsequent quotes, until noted, are all from this source.
6. Barbara Frum, *As It Happened*, McClelland and Stewart, 1976, p. 10.
7. *The CBC Radio Revolution*, p. 126.
8. *Turning Points*, p. 7.

9. *The CBC Radio Revolution*, p. 24. The Meggs-Ward report, or the English Radio Report—to give it its proper name—has never been made public by the CBC, though copies of it are extant. Legend, when I first worked at the CBC, attributed extraordinary influence to this document, as if it had been virtually the blueprint for "information radio" and for the radio revolution generally. Clay Carter's history shows that this was not true, without denying that Meggs and Ward advanced the change that was already underway.
10. Richard Liskeard, "Mediacrities," *The Last Post*, May 1974, Vol. 4, #1.
11. In 1972, burglars broke into the headquarters of the Democratic National Committee in the Watergate complex in Washington, D.C. and planted wiretaps. The perpetrators were linked to the Republic Party and ultimately to President Richard Nixon, leading to Nixon's resignation under threat of impeachment.
12. His literary executor Taylor Stoehr collected and edited his writings on media as *Format and Anxiety: Paul Goodman Critiques the Media* (Autonomedia, 1995).
13. See Stoehr's introduction in *Format and Anxiety*, p. 19.
14. *The Tower of Babble*, p. 289.
15. This document has not been published, so far as I know. I am quoting from the copy I received at the time as a freelancer. The block capitals in subsequent quotes are in the original.
16. *The Tower of Babble*, p. 24.
17. This sentence appears at the end of the preface to his *Philosophy of Right*.
18. Judy Lamarsh, Laurier Lapierre, Michael Enright, Don Harron, and several others took turns in the host's chair before Peter Gzowski returned and the *Morningside* of beloved memory came into existence.
19. *Turning Points*, p. 14.
20. Frame told this story to historian Clay Carter. See Carter, p. 117.
21. Benedict Anderson, *Imagined Communities: Reflections on the Origin and Spread of Nationalism*, Verso, 1983.
22. Bronwyn Drainie, "Once more into the fray: time for change at CBC Radio," *The Globe and Mail*, February 3, 1990, C3.
23. Sarah Lysecki, "Twists and Turns," *Ryerson Review of Journalism*, Summer, 2003.
24. *The CBC Radio Revolution*, p. 60.
25. "CBC Is Losing Sight of Its Objectives," *The Globe and Mail*, February 5, 1977.
26. Blaik Kirby, "Producers Right—CBC Radio Is Empty," *The Globe and Mail*, January 31, 1977.
27. Dennis Braithwaite, "Radio Standards Reach TV Level," *The Toronto Star*—my clipping is undated but must have been written in January 1977.
28. Reeves was a novelist, composer, and long-distance runner as well as a radio producer. Many of the themes of the producers' revolt are satirically treated in his novel *Murder by Microphone* (Avon, 1980).
29. One summit of this view of society as a system was American sociologist Talcott Parsons' *The Social System* (Routledge, 1951/1991).

Chapter 4

1. Bronwyn Drainie, "Once More Into the Fray: Time for Change at CBC Radio," *The Globe and Mail*, February 3, 1990, C3.
2. Tom Peters, *Thriving on Chaos: A Handbook for a Management Revolution*, New York: Knopf, 1987, p. 45.
3. The phrase comes from an article called "Magna Carta for the Knowledge Age." It was coauthored by Esther Dyson, George Gilder, and Jay Keyworth and appeared first in *New Perspectives Quarterly* (Fall, 1994).
4. https://www.niagarainstitute.com/.
5. Ibid.
6. *Ivan Illich in Conversation*, p. 253.
7. These thinkers will be examined more closely in Chapter Eight.
8. Sarah Lysecki, "Twists and Turns," *Ryerson Review of Journalism*, Summer, 2003.

Chapter 5

1. *George Grant in Conversation*, p. 186.
2. I am quoting Ivan Illich's summary of the Council's teaching, which drew on the writings of John of Damascus. See David Cayley, *The Rivers North of the Future: The Testament of Ivan Illich*, House of Anansi, 2005, p. 114. This definition quelled a raging ecclesiastical debate about the status of images which had at times broken out in open war.
3. https://www.cbc.ca/radio/ideas/remembering-ken-haslam-1.3889319.
4. Roland Barthes, *Mythologies*, Paladin, 1972, p. 131.
5. *Plastic Words*, p. 23.
6. Andrew Potter, Joseph Heath, *The Rebel Sell: Why the Culture Can't Be Jammed*, HarperCollins, 2004.
7. Quoted in Frank Zingrone, *The Media Simplex: At the Edge of Meaning in an Age of Chaos*, Stoddard, 2004, p. 8.
8. Ivan Illich, *Limits to Medicine: Medical Nemesis: The Expropriation of Health*, London: Marion Boyars, 1976, p. 8.
9. C. G. Jung, *Psychological Types* (Collected Works, Vol. 6), Princeton University Press, 1971, p. 426.

Chapter 6

1. *Passion and Conviction*, p. 38.
2. See Chapter One, p. 19.
3. *Turning Points*, p. 5.
4. Herschel Hardin, *A Nation Unaware: The Canadian Economic Culture*, Vancouver: J. J. Douglas, 1974, p. 140.
5. *The Politics of Canadian Broadcasting 1920–1951*, p. 413.

6. *Turning Points*, p. 37.
7. Plato presents the idea of a guardian class in Books V and VI of his *Republic.* Among others, Jane Jacobs has developed a modern version of the idea in her *Systems of Survival* (New York: Vintage Books, 1992).
8. *Inside Seven Days*, p. 180
9. The story of Watson's proposing himself for CBC President also comes from Koch, but Watson himself corroborates it in his autobiography, *This Hour Has Seven Decades* Toronto: McArthur and Co., 2005.
10. Peers, *The Public Eye*, p. 394.
11. In 1976, the CRTC's scope was expanded to include telecommunication, and its name changed to the Canadian Radio-television and Telecommunications Commission, preserving the acronym.
12. *The Public Eye*, p. 408.
13. Ibid., p. 407.
14. Ibid., pp. 160–161.
15. David Cayley, *Northrop Frye in Conversation*, Toronto: House of Anansi, 1992, p. 140.
16. "Nationalist heroes" is a quote from Herschel Hardin, a historian of the CRTC (Vancouver: Douglas and McIntyre, 1985). See *Turning Points*, p. 41.
17. *Turning Points*, pp. 41–42.
18. Monica MacDonald, *Recasting History: How CBC Television Has Shaped Canada's Past*, McGill-Queen's University Press, p. 67.
19. *Turning Points*, p. 44; Hugh Gauntlett joined CBC Television in 1955 and retired as head of Arts, Music and Science in 1988.
20. *Turning Points*, p. 44.
21. *The Public Eye*, pp. 47–49.
22. Ian Urquhart, "Ottawa: On Guard for Thee etc.," *Maclean's*, March 7, 1977; subsequent quotations are also drawn from this article.
23. David Cayley, "MPs and media missed the point of the CRTC report," *The Vancouver Sun*, August 18, 1977, p. 6; subsequent quotes from the Boyle committee's report are from this op-ed, my attempt to free the report from the political frameworks which led to its being misrepresented and ignored.
24. See previous note.
25. Karl Marx, *The Eighteenth Brumaire of Louis Bonaparte*, Mondial, 2005, p. 1.
26. *Turning Points*, p. 45.
27. *Inside Seven Days*, p. 20—the offending show was part of an explicitly experimental series called *Quest*—"a free form exercise in the inventive use of television," according to its executive producer Ross McLean—see https://en.wikipedia.org/wiki/Quest_(Canadian_TV_series).
28. Frank Peers reports extensively on these outbursts in his two volumes on the history of the CBC, *The Politics of Canadian Broadcasting* and *The Public Eye.*
29. Knowlton Nash, *The Microphone Wars: A History of Triumph and Betrayal at the CBC*, Toronto: McClelland and Stewart, 1994, p. 475.
30. Ibid., p. 434.
31. "Cutting the CBC: 11 Stations and 160 Programs Die in a Severe Corporation Budget Squeeze," *Maclean's*, December 17, 1990.
32. *Fade to Black*, p. 2, vii.
33. *Fade to Black*, p. 223.

34. *This Hour Has Seven Decades*, p. 583ff.
35. Gayle MacDonald, "CBC Staffers Shocked as Outsider Gets Top Job," *The Globe and Mail*, July 22, 2004.
36. Ibid.
37. *The Tower of Babble*, p. 18.
38. Ibid., p. 11.
39. *The Tower of Babble*, pp. 218–219.
40. Ibid., p. 241.
41. Ibid., p. 79.
42. Etan Viessing, "CBC chief defends Richard Stursberg firing," *The Hollywood Reporter*, August 12, 2010.
43. Amybeth McNulty as Anne, Geraldine James as Marilla, and R. H. Thompson as Matthew.
44. Aristotle, *Nicomachean Ethics*, Book VI, sect. 2, 1139b.
45. *The Tower of Babble*, p. 284.
46. Johnson used the phrase in testimony to the CRTC in 1978. See my "Who's Watching? Who's Listening? Who Cares? A Report on the CRTC Hearings, *This Magazine*, Vol. 12. #'s 5&6, December 1978, pp. 39–42.
47. The expression "licentious republic" was used by the first generations of Canadian Loyalists to give voice to their sense that the United States had detached freedom from the common good—with disastrous results. See S. F. Wise and Robert Craig Brown, *Canada Views the United States: Nineteenth-Century Political Attitudes*, Macmillan, 1972.
48. The Watkins report was the product of the Task Force on Foreign Ownership and the Structure of Canadian Investment, chaired by University of Toronto political economist Mel Watkins and charged with investigating the impact of growing American control of the Canadian economy. It recommended strict regulation of foreign investment in Canada.
49. The phrase recurs in Grant's work. See particularly the essay "Canadian Fate and Imperialism," in *Technology and Empire*, Toronto: House of Anansi, 1969.
50. Frye posed this question in his Conclusion to Carl F. Klink's *The Literary History of Canada*, Toronto: University of Toronto Press, 1965.
51. My cousin, Bob Dale, a Mosquito pilot, had flown the reconnaissance for a bombing raid on Nuremberg that was discussed in *The Valour and the Horror*. He was distressed by the way he was portrayed and became one of the leaders in the campaign against the series.
52. Marc Raboy, *Missed Opportunities: The Story of Canada's Broadcasting Policy*, McGill-Queen's, 1990.

Epilogue to Part One

1. John Stuart Mill, *On Liberty*, The Floating Press, 2009/1859, p. 34.
2. These words are inscribed on the Albert Einstein monument on the grounds of the National Academy of Sciences in Washington, D.C.
3. https://unherd.com/thepost/coming-up-epidemiologist-prof-johan-giesecke-shares-lessons-from-sweden/.

4. https://www.statnews.com/2020/03/17/a-fiasco-in-the-making-as-the-coronavirus-pandemic-takes-hold-we-are-making-decisions-without-reliable-data/.
5. https://off-guardian.org/2020/03/17/listen-cbc-radio-cuts-off-expert-when-he-questions-covid19-narrative/; Dr. Kettner wasn't exactly "cut off," in my view, but this story does provide audio and a transcript.
6. "la plus grande crise de santé publique de son histoire"—statement in front of the prime minister's residence on March 25, 2020—https://www.youtube.com/watch?v=NzRw-AIeNuY.
7. "Forget Politics. It's Time to Fight COVID-19," *The Globe and Mail*, September 21, 2020, A12.
8. https://iris.who.int/bitstream/handle/10665/340124/PMC7947934.pdf?sequence=1&isAllowed=y.
9. https://ocla.ca/wp-content/uploads/2020/04/Rancourt-Masks-dont-work-review-science-re-COVID19-policy.pdf.
10. https://www.cochranelibrary.com/cdsr/doi/10.1002/14651858.CD006207.pub6/full.
11. https://healthydebate.ca/2020/07/topic/an-open-letter-to-pm-covid19/.
12. https://gbdeclaration.org/.
13. Collin's e-mail was obtained by the American Institute for Economic Research (AIER), which hosted the Great Barrington Declaration, through a Freedom of Information request. The full story is told here: https://www.aier.org/article/fauci-emails-and-some-alleged-science/.
14. https://www.trishwoodpodcast.com/podcast/episode-116-marianne-klowak; see also https://www.theepochtimes.com/former-cbc-journalist-tells-national-citizens-inquiry-that-national-broadcaster-censored-stories-on-covid-19-vaccine-harms_5275498.html.
15. https://www.bbc.com/mediacentre/2020/trusted-news-initiative-vaccine-disinformation.
16. https://www.nbcnews.com/news/us-news/youtube-pulls-florida-governor-s-video-says-his-panel-spread-n1263635.
17. https://www.racket.news/p/twitter-files-thread-the-spies-who?utm_source=substack&utm_medium=email.
18. https://unherd.com/2022/12/what-i-discovered-at-twitter-hq/.
19. Rodney Palmer has worked for CBC News and has been a bureau chief for CTV news. He presented his analysis of the CBC's pandemic coverage to the National Citizen's Inquiry, a citizen-initiated and citizen-funded inquiry into COVID-19 policy—https://rumble.com/v48xctk-nci-witness-testimony-re-broadcast-rodney-palmer-mar-30-2023-toronto-ontari.html. See also https://www.science-upfirst.com/.
20. https://citizenshearing.ca/.

Chapter 7

1. Stephen Marche's 2022 book *The Next Civil War: Dispatches from the American Future* (Simon and Schuster) is just one of many possible examples.
2. Quoted in Jurgen Habermas, *The Structural Transformation of the Public Sphere*, Cambridge: MIT Press, 1989, p. 99.

3. Ibid., pp. 65–66.
4. *The Spectator*, March 12, 1711.
5. *Structural Transformation*, p. 25.
6. See especially *Habermas and the Public Sphere*, ed. Craig Calhoun, MIT Press, 1993.
7. *Structural Transformation*, p. 94; Habermas is quoting Edmund Burke.
8. Ibid., pp. 215, 175.
9. *Structural Transformation*, pp. 199, 211.
10. Ibid., pp. 172, 169.
11. Ibid., p. 160.
12. Ibid., p. 165.
13. Ibid., p. 195.
14. J. S. Mill, *On Liberty*, J. W. Parker and Son, 1859, pp. 21, 224, 130.
15. Alexis de Tocqueville, *Democracy in America*, trans. G. Lawrence, Harper and Row, 1966, p. 400.
16. "The Present Age," in *A Kierkegaard Anthology*, ed. Robert Bretall, Princeton, 1946, p. 265.
17. Ibid., p. 264.
18. Walter Lippmann, *The Phantom Public*, Harcourt Brace and Co., 1925, p. 171.
19. See https://www.davidcayley.com/podcasts/category/Origins+of+Modern+Public. Among the project's publications were *Making Publics in Early Modern Europe: People, Things, Forms of Knowledge*, ed. Bronwen Wilson and Paul Yachnin, Routledge, 2010; and *Forms of Association: Making Publics in Early Modern Europe*, ed. Paul Yachnin and Marlene Eberhart, University of Massachusetts Press, 2015.
20. Habermas, obviously, wrote in German, in which he spoke of *Öffentlichkeit*, which literally means something like public-ness, or publicity, insofar as the latter word evokes a quality rather than a product. *Public sphere* was the equivalent chosen by Habermas's translator. The expression has since become a widely used term of art in English, partly owing to the enthusiastic reception of Habermas's book, so I attribute the word to Habermas.
21. Kirby speaks in the second program of "The Origins of the Modern Public," cited above.
22. Kierkegaard, p. 269.
23. Bruno Latour, *Facing Gaia: Eight Lectures on the New Climate Regime*, Polity, 2017, p. 31.
24. I have given a much more thorough account of Latour's work than I have room for here in the following essay: https://www.davidcayley.com/blog/category/Gaia.
25. *Facing Gaia*, p. 143.
26. Bruno Latour, *Politics of Nature*, Harvard, 1999, p. 245.
27. *Facing Gaia*, p. 223.
28. Ibid., p. 223.
29. Ibid., p. 223.
30. This point, made only in passing here, is taken up at more length in the final chapter where I will explore the uses of formality and critical distance.
31. Andrew Lawton, a friendly but fair-minded reporter, has provided the most comprehensive account available so far in his *The Freedom Convoy: The Inside Story of Three Weeks That Shook the World* (Sutherland House, 2022).

32. I'm quoting Bruce Pardy, a professor of law at Queen's, who was present during some of the demonstrations in Ottawa. In conversation with Julie Ponesse, he explained the ebullient and celebratory mood of the demonstrations as an effect of the happiness people were feeling at "finding one another." The full conversation is here: https://www.youtube.com/watch?v=2FuJBTcTJn0.
33. Robert Fife and Steven Chase, "Emergencies Act Needed: Security Adviser," *The Globe and Mail*, March 11, 2022, p. A3.
34. Gillian Findlay, "The Convoy and the Questions: How a Protest Paralyzed a Capital," *The Fifth Estate*, February 22, 2022.
35. January 6, 2023.
36. See, for example, the stories collected by Ian Brown in *The Globe and Mail* ("They came. They idled. They left," June 4, 2022).
37. https://www.youtube.com/watch?v=5I0tk6OO5sw

Chapter 8

1. Quoted in Jacques Ellul, *The Humiliation of the Word*, William B. Eerdmans, 1985, p. 196.
2. Eric and Marshall McLuhan, *The Laws of Media: The New Science*, University of Toronto Press, 1988, p. 239.
3. Harold Innis, *The Bias of Communication*, University of Toronto Press, 1951, p. vii.
4. Marshall McLuhan, *The Gutenberg Galaxy: The Making of Typographic Man*, University of Toronto Press, 1962, p. 158.
5. Kevin Roose, "How the Biden Administration Can Help Solve Our Reality Crisis," *New York Times*, February 2, 2021; one of the "solutions" canvassed in this article, as I mentioned in my Introduction, is the appointment of a "reality czar."
6. The point has been developed at length by John Durham Peters in his book *The Marvelous Clouds: Towards a Philosophy of Elemental Media*, University of Chicago Press, 2015.
7. Ibid., p. 4.
8. See McLuhan's introduction to the 1964 reissue of *The Bias of Communication*.
9. See, for example, Benjamin Whorf, *Language, Thought, and Reality: Selected Writings of Benjamin Lee Whorf*, ed. John B. Carroll, *MIT Press*. 1956.
10. C. G. Jung, *Collected Works, Volume 8: Structure and Dynamics of the Psyche*, Princeton: Bollingen Series XX, para. 358.
11. Ludwik Fleck, *Genesis and Development of a Scientific Fact*, transl. by Fred Bradley and Thaddeus J. Trenn, Thaddeus J. Trenn, and Robert K. Merton eds., "Foreword" by Thomas S. Kuhn, Chicago University Press 1979.
12. 1 Corinthians 13:12; *The Bias of Communication* (1951), p. vii.
13. Marshall McLuhan, *Understanding Media: The Extensions of Man*, McGraw Hill, 1964, p. 178.
14. *The Legacy of Harold Innis*, p. 16
15. *The Bias of Communication*, p. 87.
16. Ibid., p. 140.

17. Alexander John Watson, *Marginal Man: The Dark Vision of Harold Innis*, University of Toronto Press, 2006, p. 90.
18. *The Complete Poetry and Prose of William Blake*, ed. David V. Erdman, Anchor Books, 1988, p. 37.
19. Jean Baudrillard, *Selected Writings*, ed., Mark Poster, Polity Press, 2001, p. 152.
20. Ibid., p. 169.
21. Ibid., p. 189.
22. Ibid., pp. 247, 277.
23. Ibid., p. 258.
24. Ibid., p. 260.
25. *Within the Context of No Context*, pp. 3–4.
26. Harold A. Innis, *Empire and Communications*, Rowman and Littlefield, 2007, p. 196.
27. "A Plea for Time" was the title of a lecture Innis delivered at the University of New Brunswick in 1950. See Harold A. Innis, *Staples, Markets, and Cultural Change*, ed. Daniel Drache, McGill Queen's, 1995, p. 368.
28. Eric Havelock, *The Crucifixion of Intellectual Man*, Beacon Press, 1950, p. 14; Innis and Havelock were colleagues and interlocutors at the University of Toronto between 1929 and 1947.
29. *The Bias of Communication*, p. 6.
30. McLuhan refers to this story in several places in his writings. Kevin McMahon's documentary film *McLuhan's Wake* (2002) explores his identification with Poe's tale.
31. Mark Crispin Miller, *Boxed in: The Culture of TV*, Northwestern, 1988, p. 16.
32. *Laws of Media*, p. 239. *A New Science* was the subtitle of this book; it echoed McLuhan's beloved Giambattista Vico's *La Scienza Nuova* (1725), a study of the patterns underlying language, mythology, religion and the cycles of historical change.
33. David R. Olson, *The World on Paper,* Cambridge, 1996 p. 155ff.
34. *Ivan Illich in Conversation*, p. 126ff. Subsequent quotes from the same pages.
35. David Cayley, *Ivan Illich: An Intellectual Journey*, University Park: The Pennsylvania State University Press, 2021, p. 261.
36. Leo Strauss, *Liberalism Ancient and Modern*, University of Chicago Press, 1968/1995, p. 9.
37. Simone Weil, *Gravity and Grace*, Routledge and Kegan Paul, 1952, p. 132.
38. Steven Shapin, "Is There a Crisis of Truth?," *Los Angeles Review of Books*, December 2, 2019; and Jonathan Rauch, *The Constitution of Knowledge: A Defense of Truth*, Brookings Institution Press, 2021.

Chapter 9

1. *On Liberty*, pp. 82–83.
2. *Dialogues of Alfred North Whitehead*, as recorded by Lucien Price, Greenwood Press, 1977 (1954), p. 16.

3. John G. Neihardt, *Black Elk Speaks*, Pocket Books/Simon and Shuster, 1972 (1932), p. 159.
4. *The End of the CBC?*, pp. 167, 5, 23.
5. https://en.wikipedia.org/wiki/Reality-based_community.
6. https://www.bbc.com/mediacentre/2020/trusted-news-initiative-vaccine-disinformation.
7. https://en.wikipedia.org/wiki/List_of_fact-checking_websites.
8. Kevin Roose, "How the Biden Administration Can Help Solve Our Reality Crisis," *New York Times*, February 2, 2021.
9. The Board was subsequently disbanded: https://en.wikipedia.org/wiki/Disinformation_Governance_Board.
10. *The Constitution of Knowledge*, p. 26.
11. Batya Ungar-Sargon, *Bad News: How Woke Media Is Undermining Democracy*, Encounter Books, 2021, p. 193.
12. Terry Glavin, "How Narrative Replaces the Facts," *National Post*, June 2, 2022, A1.
13. Terry Glavin, "The Year of the Graves: How the World's Media Got It Wrong on Residential School Burial Sites," *National Post*, May 28, 2022.
14. https://c2cjournal.ca/2023/12/grave-error-correcting-the-false-narrative-of-canadas-missing-children/.
15. *Constitution of Knowledge*, p. 15.
16. See, for example: https://plato.stanford.edu/entries/truth-pluralist/.
17. Lorraine Daston, Peter Gallison, *Objectivity*, Zone Books, 2007, p. 52.
18. Michael Schudson, *Discovering the News: A Social History of American Newspapers*, Basic Books, 1978.
19. David Cayley, *The Politics of Information*, *Ideas*, CBC Radio May 1–21, 1983, p. 2; the four programs in this series are available as podcasts and, in transcription, on my website: https://www.davidcayley.com/transcripts and https://www.davidcayley.com/podcasts/category/Politics+of+Information. I will cite the series by transcript page.
20. *Politics of Information*, p. 4.
21. Ibid., p. 5.
22. I surveyed this scholarship, up to 1983, in *The Politics of Information*.
23. The war for independence from France began in 1946; the last American troops withdrew in 1973.
24. https://www.cbc.ca/news/world/the-fall-of-saigon-how-cbc-ctv-covered-the-1975-events-1.3055237.
25. A fuller account of these events in given in the epilogue that ends the first part of the book.
26. Christopher Clark, *The Sleepwalkers: How Europe Went to War in 1914*, HarperCollins, 213, p. 562.
27. "Forget Politics. It's Time to Fight COVID-19," *The Globe and Mail*, September 21, 2020, A12.
28. Noam Chomsky and Edward S. Herman, *The Political Economy of Human Rights*, Volume I, *The Washington Connection and Third World Fascism*, Volume II, *After the Cataclysm: Postwar Indochina and the Reconstruction of Imperial Ideology*, Black Rose Books, 1979.
29. The report was published as a book by the New York University Press in 1975, the same year it was submitted to the Trilateral Commission. The Commission was an

elite American NGO and power behind the throne, which was then playing a role rather similar to the one which many today ascribe to the World Economic Forum.
30. *Four Quartets*, p. 4.
31. Octavio Paz, "Reflections: Mexico and the United States, *The New Yorker*, September 17, 1979, p. 153.
32. "The idols of the tribe" was Francis Bacon's term for humanity's naïve faith that it could discern the truth with its native equipment. See Francis Bacon *Novum Organum*, Open Court, 1994, p. 54ff.
33. John Ioannidis, How the Pandemic Is Changing the Norms of Science, *Tablet*, September 8, 2021, https://www.tabletmag.com/sections/science/articles/pandemic-science.
34. Daszak's story is carefully told here. https://www.science.org/content/article/we-ve-done-nothing-wrong-ecohealth-leader-fights-charges-his-research-helped-spark-covid-19.
35. Nicholas Wade, "The Origin of COVID: Did People or Nature Open Pandora's Box at Wuhan?" *Bulletin of the Atomic Scientists*, May 5, 2021.
36. Jeremy Farrar, *Spike: The Virus v. the People: The Inside Story*, London Profile Books, 2021.
37. This and subsequent quotations are taken from Jeffrey Tucker's review of the book here: https://brownstone.org/articles/the-lab-leak-the-plots-and-schemes-of-jeremy-farrar-anthony-fauci-and-francis-collins/.
38. Offit described this meeting in an interview with fellow physician Zubin Damania who podcasts by the handle ZDoggMD: https://www.youtube.com/watch?v=wkz1ln5AJ5Q.
39. Shoshana Zuboff, *The Age of Surveillance Capitalism: The Fight for a Human Future at the New Frontier of Power*, Public Affairs, 2019, p. 352.
40. Ibid., pp. 234, 261.
41. Ibid., p. 261.
42. Joseph Bernstein, "Bad News: Selling the Story of Disinformation," *Harper's Magazine*, September 2021, p. 29.
43. https://twitter.com/jamesmelville/status/1488472558425063430?lang=en.
44. Many contemporary historians and philosophers of science have made this point. Nobel laureate Ilya Prigogine and philosopher Isabelle Stengers say, for example, that "science is … the prophetic announcement of a description of the world seen from a divine … point of view." (*Order Out of Chaos: Man's New Dialogue with Nature*, New Science Library, Shambhala Books, Boulder, p. 76) Bruno Latour has explored the ways in which the "modern constitution" "crossed out" God and ascribed God's attributes to Nature, as interpreted by science. (See particularly *We Have Never Been Modern* and *Facing Gaia*.)
45. https://nationalpost.com/opinion/bruce-pardy-how-canadas-secular-religion-of-cultural-self-hate-took-hold.
46. Thomas King, *The Truth About Stories: A Native Narrative*, Toronto: House of Anansi, 2003, p. 2.
47. Wallace Stevens, *Collected Poetry and Prose*, Library of America, 1997, p. 331.
48. Revelation, 21:2.
49. Rene Descartes, *A Discourse on Method*, Oxford World's Classics, 2006, p. 51; Friedrich Engels, *Anti-Dühring*, Foreign Languages Publishing House, 1959, p. 387.
50. *Facing Gaia*, p. 223.

51. See Connolly's chapter in *Agonistic Democracy*, ed. Mark Wenman, Cambridge, 2013; Chantal Mouffe, *Agonistics: Thinking the World Politically,* Verso, 2013; and https://en.wikipedia.org/wiki/Agonism
52. *Agonistics*, p. 74.
53. https://cbc.radio-canada.ca/en/vision/governance/journalistic-standards-and-practices.
54. *Truth and Method*, p. 361.
55. *The Way of Life According to Lao Tzu*, p. 66.

Chapter 10

1. *Truth and Method*, p. 299.
2. Franz Rosenzweig, *Philosophical and Theological Writings*, translated and edited by Paul W. Franks and Michael L. Morgan, Indianapolis/Cambridge: Hackett Publishing. 2000, p. 126.
3. Emma Jung and Marie-Louise von Fran, *The Grail Legend*, Princeton University Press, 1998, p. 294.
4. Robert W. McChesney, "Graham Spry and the Future of Public Broadcasting: The 1997 Spry Memorial Lecture," *Canadian Journal of Communication*, Volume 24 Issue 1, January 1999; McChesney cites: Canada, House of Commons, 1932, p. 546.
5. See Chapter One, p. 5.
6. https://cha-shc.ca/advocacy/the-history-of-violence-against-indigenous-peoples-fully-warrants-the-use-of-the-word-genocide/; https://torontosun.com/video/6a1295ac-bd0c-11ee-9d02-b6161e45561f/dr-jordan-peterson-explains-canada-is-a-sinking-ship.
7. *Truth and Method*, p. xxviii.
8. Ibid, p. 385.
9. *This Hour Has Seven Decades*, p. 250.
10. Ibid., p. 570.
11. Ibid., p. 562.
12. Ibid., p. 250.
13. Ibid., p. 133.
14. https://www.cbc.ca/news/canada/saskatchewan/lawyer-says-residential-school-denialism-should-be-added-to-criminal-code-1.6882579.
15. R. G. Collingwood, *An Autobiography and Other Writings*, eds. David Boucher and Teresa Smith, Oxford University Press, 2013, p. 100.
16. James Carse, *Finite and Infinite Games: A Vision of Life as Play and Possibility*, Random House (Ballantine Paperback edition), 1986, pp. 22–23.
17. "And we are here as on a darkling plain/ Swept with confused alarms of struggle and flight,/ Where ignorant armies clash by night." Matthew Arnold, Dover Beach.
18. Many of the discourses of decolonization now suppose that this tradition can be renounced, overcome and set aside. It often appears to me that these arguments are made from within Western assumptions—that what is ostensibly being cast off is what is also doing the casting off—but for the purposes of this book I'll leave the point moot.

19. "Epistemological Crises, Dramatic Narrative, and the History of Science" in Alasdair MacIntyre, *The Tasks of Philosophy: Selected Essays, Vol 1*, Cambridge University Press, 2006.
20. Ibid., p. 12.
21. Ibid., p. 5.
22. *An Autobiography*, p. 58.
23. Christopher Dummitt and Zachary Patterson, *The Viewpoint Diversity Crisis at Canadian Universities: Political Homogeneity, Self-Censorship and Threats to Academic Freedom*, Ottawa: McDonald-Laurier Institute, 2022, pp. 4–5.
24. See p. 126.
25. Dave Snow, "The CBC Prioritizes Allyship over Objectivity in Saskatchewan Parental Consent Coverage: An Empirical Analysis," *The Hub*, December 20, 2023; Tara Henley, "The Trust Spiral: Restoring Faith in the Media," *Literary Review of Canada*, May, 2024.
26. https://laws-lois.justice.gc.ca/eng/acts/b-9.01/FullText.html.
27. Robert E. Babe, *Life Is Information: The Communication Thought of Graham Spry* in *Media, Structures and Power*, ed. Edward A. Comor, University of Toronto Press, 2011, pp. 65–66.
28. Chapter Two, pp. 36
29. *Truth and Method*, p. 357.

Chapter 11

1. "Letter on Humanism," in Martin Heidegger, *Basic Writings*, New York: HarperCollins, p. 193.
2. *Truth and Method*, p. 440.
3. https://arshabodha.org/wp-content/uploads/abc/teachings/Kathopanishad/kathaTrans1.pdf.
4. Genesis 2:19.
5. Friedrich Nietzsche, *Untimely Meditations*, Cambridge University Press, 1997, p. 214.
6. *Four Quartets*, pp. 7, 16.
7. "The Circus Animals' Desertion" in *The Collected Poems of William Butler Yeats*, Wordsworth Editions, 2000, p. 297; "The Wasteland" in Thomas Stearns Eliot, *Collected Poems 1909–1962*, Harcourt, Brace, and World, 1963.
8. Harold A. Innis, *Changing Concepts of Time*, Rowman and Littlefield, 2004, p. 85.
9. *Commodify Your Dissent Salvos from the Baffler*, eds. Thomas Frank and Matt Weiland, W. W. Norton, 1997.
10. *Ivan Illich in Conversation*, p. 213.
11. Richard Dawkins, *The Selfish Gene*, Oxford, 2016, p. 246.
12. *Novum Organum*, p. 55, 64.
13. Thomas Spratt, *The History of the Royal Society*, S. Chapman, 1722, p. 113.
14. Ibid—from a prefatory poem by Thomas Cowley which is unpaginated; and *Novum Organum*, p. 140.
15. Charles Taylor, *The Language Animal: The Full Shape of the Human Linguistic Capacity*, Harvard, 2016.

16. *Plastic Words*, p. 91ff.
17. *Truth and Method*, p. 454.
18. Pörksen is a friend, and I am listed as one of the translators of his already cited book *Plastic Words*, though, in truth, my German-born wife Jutta Mason did all the heavy lifting, and I only helped to refine the book's style in English.
19. *Ivan Illich in Conversation,* p. 253. In his *Deschooling Society* (Penguin, 1973, p. 32) Illich suggested that certain words like "amoebas" have become so flexible that they "fit into almost any interstice of the language." This seed grew into the idea of plastic words in conversations between Illich and Pörksen when they were both fellows at the *Wissenschaftskolleg* in Berlin in 1980. See my *Ivan Illich: An Intellectual Journey*, pp. 279–281.
20. This was said in an *Ideas* program I did with Pörksen. See *Plastic Words*, p. 6, here: www.davidcayley.com/transcripts.
21. I hope it will be understood that I am talking here about the fate of words and not about the validity of Selye's scientific theory of how glucocorticoids generate the "stress response."
22. This research was carried out by Barbara Duden and Silja Samerski and is described by Barbara Duden in *Ideas on the Nature of Science*, ed. David Cayley, Goose Lane, 2009, pp. 259–262.
23. A. N. Whitehead, *Science in the Modern World*, Mentor Books, 19490/1925, p. 52.
24. Source code specifies various functions, definitions, and methods that are to be *executed* by a computer.
25. Eric Andrew Gee and Tavia Grant, "In the Dark: The Cost of Canada's Data Deficit," *The Globe and Mail*, January 26, 2019.
26. Erwin Schrödinger, *What Is Life? The Physical Aspect of the Living Cell*, Cambridge University Press, 1944.
27. Norbert Wiener, *The Human Use of Human Beings: Cybernetics and Society*, Hachette Books, 1988/1954, p. 96.
28. Merlin Sheldrake, *Entangled Life: How Fungi Make Our Worlds, Change Our Minds and Change Our Futures*, Toronto: Penguin Random House, 2021.
29. René Descartes, *Rules for the Direction of the Mind*, Rule IV—*There is need of a method for finding out the truth*—paragraphs four and five. The text is here: http://dt.pepperdine.edu/descartes-rules-for-direction-of-the-mind.html.
30. See note 15.
31. Clark Whelton, "What Happens in Vagueness Stays in Vagueness: The Decline and Fall of American English and Stuff," *City Magazine*, Winter, 2011.

Chapter 12

1. Hannah Arendt, *The Life of the Mind, Volume One: Thinking*, Harcourt, Brace, Jovanovich, 1971, p. 178.
2. Stefan Müller-Doohm, *Adorno: A Biography*, Polity Press, 2005, pp. 440–441.
3. Ibid., p. 475.
4. Ibid., p. 463.
5. https://en.wikipedia.org/wiki/Theodor_W._Adorno#Confrontations_with_students.

6. Ibid.
7. *Adorno: A Biography*, p. 464.
8. Theodor Adorno, *The Culture Industry: Selected Essays on Mass Culture*, ed. J. M. Bernstein, Routledge, 1991, pp. 198–203.
9. Adorno ascribes the quoted phrase to his colleague Jurgen Habermas.
10. In the edition I have cited, the translation is, "Only thinking could *offer an escape*." The alternative I have used is here: http://www.autodidactproject.org/quote/adornoresgn.html.
11. *The Legacy of Harold Innis*, p. 8.
12. I heard this remark from economist Mel Watkins who quotes it here: https://rabble.ca/blogs/bloggers/mel-watkins/2014/09/harold-innis-goes-to-war.
13. See, for example, Mel Watkins's comments in *The Legacy of Harold Innis*, p. 8.
14. Rick Salutin quotes this remark of Innis's in a column he wrote for the *Globe and Mail* on September 7, 2007, called "The Upside of Ivory Towers."
15. *The Bias of Communication*, p. 11.
16. The story of Dr. Schabas's exclusion from the CBC is told, and documented, in the Epilogue to Part One, p. 123ff.
17. Conservative Party of Canada Policy Declaration, September 9, 2023, p. 35, https://cpcassets.conservative.ca/wp-content/uploads/2023/11/23175001/990863517f7a575.pdf; Conservative leader Pierre Poilièvre has at times spoken even more bluntly about simply "defunding" the CBC, and his party has initiated an official petition entitled "Defund the CBC," https://www.conservative.ca/cpc/defund-the-cbc/.
18. Campaigns along these lines in the 1940s and 1950s are described in Chapter One, p. 23ff.
19. See, for example, Mike Lapointe, "'Tone and Tenor' of Calls to Defund the CBC Different This Time Around, Say Media Experts, as Conservatives Hammer Away at the Public Broadcaster," *The Hill Times*, May 15, 2024.
20. "Political Monoliths"—see p. 186.
21. See p. 152ff.
22. *The Life of the Mind*, p. 88.
23. Ibid., p. 178.
24. Theodor W. Adorno, *Negative Dialectics*, Continuum, 1973, p. 365.
25. Ibid., p. 314.
26. *The Life of the Mind*, p. 185.
27. Peter Brooks, *Seduced by Story: The Use and Abuse of Narrative*, New York Review of Books, 2022, p. 4.
28. Milan Kundera, *The Unbearable Lightness of Being*, Harper and Row, 1984, p. 251.
29. King, p. 2.
30. Ronald Niezen, *Truth and Indignation: Canada's Truth and Reconciliation Commission on Indian Residential Schools*, University of Toronto Press, 2013, p. 84.
31. Ibid., p. 89.
32. *Truth and Indignation*, p. 91.
33. https://www.cbc.ca/news/politics/residential-school-system-well-intentioned-conservative-senator-1.4015115; https://www.cbc.ca/news/politics/lynn-beyak-stands-by-fake-news-1.4028126.
34. The section of *Summa Theologica* concerning heresy is here: https://www.newadvent.org/summa/3011.htm.

35. "Takedown" refers to the attack on the authors of the Great Barrington Declaration that was initiated by the head of the American National Institutes of Health (NIH) in 2021. The story is told in the Epilogue to Part One, p. 125.
36. Hal Niedzviecki, *Write*, Writer's Union of Canada, Spring, 2017.
37. https://en.wikipedia.org/wiki/Hal_Niedzviecki; D. A. Lockhart, a member of the Moravian of the Thames First Nation in Chatham-Kent, reacted to Niedzviecki's statement by telling the CBC: "We've lost our land, we've lost our languages and almost the last thing we have left are our stories and our voices. To have somebody come in and say we'll tell those better than you … is sort of a painful kick while you're already down." https://www.cbc.ca/news/canada/windsor/appropriation-prize-a-kick-while-you-re-already-down-says-indigenous-author-1.4116494.
38. At the beginning of the pandemic, it was widely reported that the infection mortality rate might be as high as 3 percent. It turned out to be closer to 0.25 percent, falling to 0.05 percent when those over seventy were excluded. See the Epilogue to Part One, p. 122 for reference.
39. *The Legacy of Harold Innis*, p. 8.
40. See Blake's poem "London" in *Complete Poetry and Prose*, p. 27.
41. Letter to George and Tom Keats, December 21 (27?), 1817 in John Keats, *Selected Letters*, Oxford University Press, 2002, pp. 41–42.
42. *Gravity and Grace*, p. 93.
43. *Dialogues of Alfred North Whitehead*, p. 16.
44. *Life of the Mind*, p. 185.

Chapter 13

1. *Liberalism Ancient and Modern*, p. 9.
2. *Habermas and the Public Sphere*, p. 377.
3. Paolo Freire, *The Pedagogy of the Oppressed*, Bloomsbury Publishing, 2018, p. 67 (first published in English in 1970).
4. Ibid., pp. 88, 77, 135.
5. Ibid., p. 89.
6. Henry David Thoreau, *Walden*, New York: Thomas Y. Crowell and Company, 1910, p. 122.
7. This saying has sometimes been traced back to a 1970 essay by feminist Carol Hanisch which used it as a title, but by then it was already a slogan in wide use.
8. Richard Sennett, *The Fall of Public Man*, Cambridge, 1974, p. 296.
9. Ibid., p. 4.
10. See Chapter Three for examples of this emphasis on format and on communion with audiences.
11. Hannah Arendt, *The Human Condition*, Chicago, 1958, p. 199ff.
12. Grant appeared regularly on the CBC throughout his career, and two of his books, *Philosophy in the Mass Age* and *Time as History* began life as lecture series for the CBC. I was honoured to present his last broadcast on *Ideas*, two years before he died—https://www.davidcayley.com/podcasts/category/George+Grant—in a documentary series called *George Grant: The Moving Image of Eternity.* The long interview on which this series was based was later published as *George Grant in Conversation.*

13. *Ideas*, CBC Radio, June 6, 2023.
14. See Chapter Eight, p. 153, for a discussion of rhetoric's place in the old university curriculum.
15. See Walter Ong's *Ramus, Method, and the Decay of Dialogue: From the Art of Discourse to the Art of Reason* (Harvard University Press, 1958); and *Rhetoric, Romance and Technology* (Cornell, 1971).
16. *Rhetoric, Romance and Technology*, p. 8; subsequent quotations on the same page.
17. *Structural Transformation*, p. 25.
18. *The Human Condition*, p. 50.
19. Ibid, pp. 64, 199.
20. Nina Eliasoph, *Avoiding Politics: How Americans Produce Apathy in Everyday Life*, Cambridge University Press, 1998, p. 255.
21. Ibid., p. 255.
22. Michael Warner, *Public and Counterpublics*, Zone Books, 2002, p. 108.
23. *Habermas and the Public Sphere*, p. 377.
24. *Publics and Counterpublics*, p. 165; subsequent quotes, until noted, are from the same page.
25. *ab-[s]tractus*—drawn away from.
26. *Publics and Counterpublics*, p. 160.
27. Linguists, since Nikolay Trubetzkoy (1890–1938), distinguish between marked and unmarked terms. In the pair man/woman, for example, *man* is unmarked, since it is commonly used to refer to everyone, while *woman*, as referring only to females, is marked.
28. Charles Taylor, "The Politics of Recognition," in *Multiculturalism: Examining the Politics of Recognition*, ed. Amy Gutmann, Princeton University Press, 1994; Taylor and Warner have been colleagues and interlocutors.
29. *Publics and Counterpublics*, p. 160.
30. Ibid., p. 168.

Index